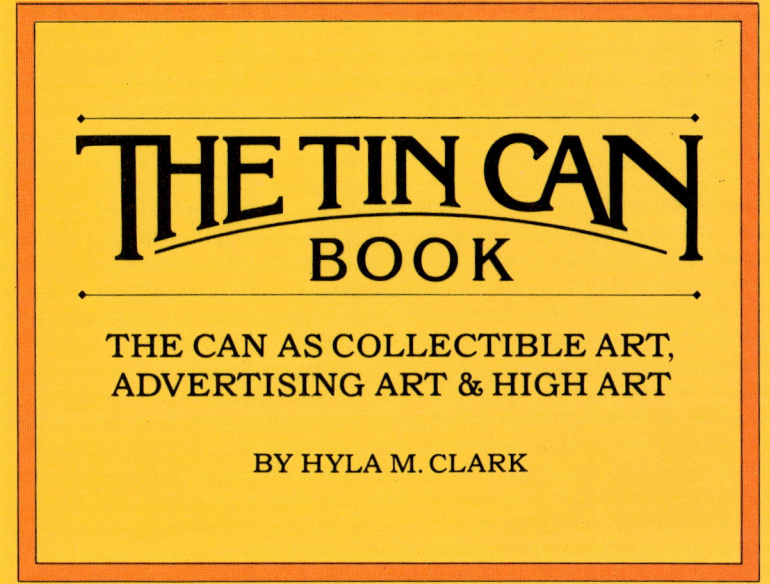

THE TIN CAN BOOK

THE CAN AS COLLECTIBLE ART, ADVERTISING ART & HIGH ART

BY HYLA M. CLARK

NEW YORK, LONDON, AND SCARBOROUGH, ONTARIO

Copyright © 1977 Tree Communications, Inc. All rights reserved.
No part of this work may be reproduced or transmitted in any form by any means, electronic or mechanical, including photocopying and recording, or by any information storage or retrieval system, without permission in writing from the copyright holder. Distributed by The New American Library, Inc. NAL books are also available at discounts in bulk quantity for industrial or sales-promotional use. For details, write to Premium Marketing Division, New American Library, Inc.

Library of Congress Catalog Card Number: 76-52265

SIGNET, SIGNET CLASSICS, MENTOR, PLUME and MERIDIAN BOOKS are published *in the United States* by The New American Library, Inc. 1301 Avenue of the Americas, New York, New York 10019, *in Canada* by The New American Library of Canada Limited, 81 Mack Avenue, Scarborough, 704, Ontario, *in the United Kingdom* by The New English Library Limited, Barnard's Inn, Holborn, London, E.C. 1, England.

First NAL printing: March, 1977; 1 2 3 4 5 6 7 8 9; Printed and bound in the United States of America.

Created and Produced by **Tree Communications, Inc.**, 250 Park Avenue South, New York, New York.
Publisher: **Bruce Michel**; Editorial director: **Rodney Friedman**; Design director: **Ronald Gross**; Director of photography: **Paul Levin**; Art director: **Sonja Douglas**; Production: **Lucille O'Brien**; Design assistant: **Christopher Jones**.

Assisting in the production of this book: Technical consultant: **Lester Barnett**; Text editor: **Nancy Naglin**; Copyreader: **Laurence Barandes**; Typesetter: **Iris Rautenberg**; Production assistant: **Elizabeth Henley**.

Photo and illustration credits: **From the collection of Lester Barnett,** photographs 1, 21; illustrations 10, 174, 175. **THE BETTMAN ARCHIVE, INC.**, New York, New York, illustration 18. **Can Manufacturers Institute, Inc.,** Washington, D.C., photographs 2, 9. **Henry Francis Du Pont Winterthur Museum Library,** Winterthur, Delaware, photograph 12. **Leo Castelli,** New York, New York, photographs 27, 28. **Lynn Matus,** illustrations 8, 24. **Steven Mays,** photographs 42, 220-231. **New York Public Library Picture Collection,** New York, New York, illustrations 3, 4, 6, 7, 10, 11. All other photographs by **Paul Levin**.

The tin containers which appear in this book have been taken from the following collections: **John Ahrens:** photographs 25, 26, 232-261. **America Hurrah N.Y.C.**; photographs 17, 165, 166, 176-185, 187-211. **Lester Barnett:** photographs 5, 14-16, 19, 20, 23, 29-31, 38, 46, 50, 51, 58, 73, 167-171, 173, 220-231. **Honey and Joe Freedman:** photographs 35, 53, 55, 56, 78-88, 127, 128. **Harry's Grocery:** photographs 22, 32-34, 37, 39-45, 47-49, 54, 57, 59, 60, 64, 65, 89-95, 98-100, 105-107, 117, 119-123, 125, 126, 130-138, 145-148, 150-154, 156-158, 163, 172. **Shirley Kirk:** photographs 36, 61-63, 66, 108, 149, 155, 159-162, 212, 213. **Joel and Kate Kopp:** photograph 186. **Dr. Hilton Read:** photographs 13, 52, 67-72, 74-77, 214-219. **Dolly and George Yanolko:** photographs 96, 97, 101-104, 109-116, 118, 124, 129, 139-144, 164.

The text face for this book is Century Expanded, set at Tree Communications, Inc. The display type is Korinna, set at Latent Lettering Co., Inc. Black and white prints were made by Portogallo, Inc. Color separations were made by National Colorgraphics, Inc. Halftones were made by International Plate Service Corp. The paper used is 70 lb. Consoweb Velvet made by Consolidated Papers, Inc. and supplied by Lindenmeyr Paper Corporation. The book was printed and bound by Rand McNally & Company.

CONTENTS

The Tin Can in History **5**
Chronological History of Tin Cans **11**
Evaluating Tin Containers **26** / Selected Bibliography **36**
Store Bins **37** / Coffee and Tea Containers **45**
Cocoa and Spice Containers **50**
Peanut Tins and Peanut Butter Pails **56**
English Biscuit Tins **61** / Vegetable Tins **68**
Other Food Containers **73** / Talcum Powder Tins **78**
Cosmetic and Pharmaceutical Containers **84**
Tobacco Containers **90** / Roly Poly Tobacco Tins **103**
Sample Tins **107** / Story Tins **113**
Beer Cans **119**

The author wishes to thank the tin-container collectors who graciously allowed their collections to be photographed: John Ahrens, America Hurrah N.Y.C., Lester Barnett, Honey and Joe Freedman, Harry's Grocery, Shirley Kirk, Joel and Kate Kopp, Dr. Hilton Read, and George and Dolly Yanolko. Their cooperation and hospitality made the work a pleasure. Special thanks must also be given to Lester Barnett for his inspiration and technical assistance.

1. Elegant graphics on product labels helped to attract consumer attention in the crowded environment of the country store. Colorful tins had the added advantage of decorating the store.

The story of the ubiquitous tin can is not as simple as one might expect of such a common object. It is a complicated tale, spanning several centuries and involving simultaneous histories. A tin can is not made from tin; the popular name is a contraction of the original term—tinplated canister. Tinplated canister derives from tinplated iron sheets (later steel) of which tin cans are made. Therefore, the growth of the tinplate industry, as well as the development of steel, are crucial to the history of the tin can. The mechanization of the tin container manufacturing industry itself is part of the story of the growth of industrialization. Closely related are advances in food preservation and the development of techniques for packing food in tin containers. And a discussion of the labeling of tin cans would be incomplete without a basic understanding of lithography and its application to metal printing.

The history of the tin can also tells the story of the development of retail trade and advertising. In the late eighteenth and early nineteenth centuries, when tinplated canisters first began to proliferate, peddlers brought retail goods to the consumer. By the closing decades of the nineteenth century, during the flowering of the tin can, these peddlers had become manufacturer's agents or opened country stores. Today supermarkets have replaced country stores, but the tin can, produced in ever-increasing numbers and threatened daily by synthetic materials, remains.

Tin cans and their labels have been subjects for some of this century's finest artists, but the label's original purpose was more practical. Early container labels were generally either entirely typographic or featured pictures of the product for sale. Later, when sellers became more ad-conscious and competitive, labels were not only descriptive but a potential selling aid as well.

2. Small cans, vegetable tins and store bins are all displayed on this Los Angeles Can Company wagon. During the late nineteenth and early twentieth century canning boom, can manufacturing firms opened up all over the United States.

Most of the tin cans shown in this book belong to the decorated metal box variety. The rest are vegetable tins—hermetically sealed tin containers—or beer cans, which are a twentieth century variation of the vegetable tin. Decorated metal boxes, vegetable tins, and beer cans have recently become highly collectible. At first many collectors became involved because they found the colorful containers attractive. The present popularity and availability of collectible tin cans attest to the success of the manufacturer's designs. They were able to please their original customers so successfully that their cans were preserved, sometimes to be reused but often tucked away, because they were too nice to throw out.

At present, the most collectible decorated metal boxes are from the era 1880 to 1940. (Beer can collectors begin with the first flat-top beer can made in 1935, and continue to the present day.) The tin container industry flourished throughout the given era in both Great Britain and the United States, producing remarkable cans. In the early part of the period several days were sometimes required to create an elaborately decorated metal box. The earliest cans were handmade, and even after stamping machines were in widespread use, several days' drying time were needed between applications of consecutive coats of paints, inks and lacquers.

Such enormous expenditures of time seem fabulous when compared to our high-speed technology, and to a tin collector, the names of the early manufacturers are legendary. Particular firms are especially revered. It might be Norton Brothers, which created enormous, meticulously decorated store bins in Illinois, or New York based Ginna and Company, famous for many-colored tins as well as for intricate thin-line drawings. The favorite might be Somers Brothers. The three brothers, able machinists, developed a workable lithographic process for printing on metal at

3. Bar iron is forged and hammered into sheets in an eighteenth century tinplate works. The sheets are heated in groups (see inset) and hammered again. Water wheels drive the hammers.

their plant in Brooklyn, New York. Perhaps Tindeco (Tin Decorating Company; Baltimore, Maryland), the maker of the Roly-Poly tin designed for Mayo's tobacco is preferred. (Clark Secrest writing in *Tin Type*, a monthly newsletter for tin collectors, calls these tins "the little round people.") American Can Company and Continental Can Company may even become the subjects of often-told tales.

Factories and corporations may not seem potentially heroic, but the saga of the tin can doesn't lack for individual heroes. Aloys Senefelder, Nicholas Appert, Thomas Huntley, Gail Borden, Joseph Campbell and Gerhard Mennen are some protagonists of the tale.

The story of the tin container begins with the manufacture of tinplate in fourteenth-century Bohemia. After iron bars were forged, workers hammered them by hand into sheets. When the hammering was completed, the scale was scrubbed away with an abrasive, and the descaled sheets were "pickled" in a foul-smelling brine concocted from fermenting grains. Then the sheets were rinsed and dipped in molten tin. From the tinplate, tinsmiths formed and soldered durable household goods of all descriptions.

Almost from the start tinware was a success. Early tinplate was crude—the hammered iron sheets and the tin coating were heavy and irregular—but articles made from it were inexpensive and strong. During the seventeenth century English pewterers became so alarmed by the spread of tinware that they issued a formal protest.

The tinplate, which was making successful inroads on the British market, was not a domestic product. The tinplate industry had begun in Germany, and for the next three centuries German tinplate predominated. Several attempts were made to establish tinplate works in England in the sixteenth century, but the struggling industry remained weak until the eighteenth century. German expertise was not to be as easily imported as tinplate, and the lack of skilled workers presented a nearly insurmountable problem for early English manufacturers.

Germany continued to dominate the industry until 1730. The British had issued a protective tariff in 1706, and as imported tinplate became prohibitively expensive, the domestic industry thrived. Less than two hundred years later the same situation was repeated in America with the passage of the McKinley Tariff Act in 1891.

Britain dominated the world tinplate industry from the 1730s until the end of the nineteenth century. Her flourishing tin works contained the same three basic divisions as their German predecessors. Each factory included a forge where bar iron was formed, a mill where bar iron was rolled into sheets, and a tin house where iron sheets were dipped into molten tin. Some tin works which did not forge their own iron had only two departments: a mill and a tin house. Sometimes several mills were grouped together to form one manufacturing complex, but the steps required to process tinplate remained the same. The work was heavy, the atmosphere hot and evil-smelling, and the slow process required many skilled workers.

4. Rolled steel sheets—manufactured from the 1880s until the 1920s—were more uniform than hammered sheets; but they were also handmade and far from perfect. Here workers inspect steel sheets. Those that met specifications went on to be tinned.

Almost from the very beginning water power had been used to work the hammers and by 1700, water-driven rollers were used to flatten the iron sheets. Until the middle of the nineteenth century tin works were tied to water. Significant manufacturing advancements developed slowly, and between the 1760s and the 1830s, the industry did not change substantially. But during the remainder of the nineteenth century, the production of tinplate in Great Britain increased one hundred and fifty times. The demand for tinplate increased rapidly as its primary use shifted from the production of durable household goods to the manufacture of containers for the petroleum and food preserving industries.

In the 1850s, steam power was introduced, and soon afterward two processes for making steel were invented: Bessemer steel was processed in the middle of the century and open-hearth steel in 1875. The higher quality open-hearth steel was used almost exclusively in the manufacture of tinplate from 1880 to 1900. In the twentieth century, the Bessemer process was made more economical and therefore became more popular.

The replacement of iron by steel in the manufacture of tinplate separated the bar steel manufacturer from the tinplate producer. The higher quality steel allowed a thinner tin coating than iron. Soon mechanical tinning pots, first used in 1882, made it possible to apply thinner coats of tin. Since steam power and mechanical tinning pots required less skilled labor, shifts were reduced and the number of workers declined. As both material and labor requirements decreased, tinplate became less expensive and easier to produce. Nonetheless, the steps in the tinplating process remained the same, and both large and small manufacturers were able to coexist in the expanding market.

The first American tinplate works was established in 1871. The fledgling industry faced problems similar to those which had plagued the infant British industry two centuries before. The American market for tinplate was increasing at an even greater rate than in Britain, but knowledge of the manufacturing process was limited, and there were few skilled workers available. Imported tinplate was less expensive and of higher quality than the domestic product.

The United States secured the future of its tinplate industry the same way Britain had. The establishment of a protective tariff was a hot issue in the presidential campaign of 1888. The Republicans won the election, and in 1891, the McKinley Tariff Act was passed. (McKinley was then head of the Ways and Means Committee of the House of Representatives.) The price of imported tinplate increased nearly 70 percent. At the same time, the American market for tinplate continued to expand rapidly, and by the 1920s, production in the United States had surpassed that in Great Britain.

Throughout the nineteenth century the manufacture of tinplate continued to be an independent operation. But in 1898, Daniel G. Reid, William H. Moore and William B. Leeds effected the merger of thirty-eight tinplate factories to form American Tinplate Company. Later, the consolidation became a subsidiary of United States Steel Corporation, organized in 1901.

The production of tinplate in the twentieth century was still hot and heavy work. For example, a steel bar was flattened between power-driven rollers until a pack of eight sheets, each having the surface area of the original bar, was formed. During

CHRONOLOGICAL HISTORY OF TIN CANS

DATE	HISTORICAL EVENT	LABELING METHOD
1300	Iron is tinplated in Bohemia.	handpainting
1635	English pewterers protest the spread of tinware.	
1706	The British pass a tariff protecting the tinplate industry.	
1730	Britain dominates the world tinplate industry.	
1780	A small lead drum is manufactured to contain snuff.	
1798	Aloys Senefelder invents lithography.	paper labels
1809	Nicholas Appert wins 12,000 francs for his research in food preservation.	
1810	Augustus de Heine and Peter Durand each obtain an English patent for canning in tin containers.	
1811	Bryan Donkin and John Hall obtain an English patent for canning in glass.	embossing
1819	Ezra Daggett and Thomas Kensett begin canning fish in New York, and Charles Underwood begins canning vegetables and fruits in Boston.	
1830	Huntley and Palmers package their biscuits in tin boxes.	
1847	The pendulum press is invented.	stenciling
1851	The first steam engine is placed in a tinplate works.	paper transfer
1856	Gail Borden introduces condensed milk. Bessemer steel is invented.	
1865	Colonel Silas Augustine Ilsley begins making tinware in Brooklyn, New York.	
1868	Edwin Norton begins making tin cans on a small scale.	
1869	The three Somers brothers begin making metal tags.	
1870	Barclay and Fry patent the offset press.	one-color lithography (printed on colored base)
1875	The open-hearth process for making steel is introduced.	
1876	A patent for printing directly on roughened tinplate is issued.	
1879	The Somers brothers develop their own process for lithographing tin containers.	
1882	The first mechanical tinning pots are used.	chromolithography
1890	A can is made from a single sheet of tinplate at Norton Brothers.	
1891	The United States passes the McKinley Tariff Act. Hasker and Marcuse Manufacturing Company is formed.	
1898	A patent to vacuum pack is issued to Edwin Norton. American Tinplate Company is formed. Cobb Preserving Company introduces the first fully automatic canning line.	
1900	Tindeco begins operation.	
1901	American Can Company is formed. Heekin Can opens its doors.	
1903	The rotary offset press is patented.	
1904	Sanitary Can Company is formed. Edwin Norton founds Continental Can Company.	
1914	Continuous ovens for drying inked tinplate are introduced.	
1927	Continuous hot rolling of steel comes into use.	photolithography
1935	Beer is canned.	
1937	Steel is tinplated using an electrolytic process.	
1959	Adolph Coors Co. introduces the first all-aluminum beer can.	
1962	Beer cans have aluminum tab tops.	
1965	Tin-free-steel cans are made.	

this process it was necessary to repeatedly handle the red-hot sheets. As the steel was pressed through, two men ("catchers") grabbed the sheet with tongs, ran with it, folded it in half and flipped it back over the rollers where two men stood ready to feed it through again. The workers wore heavy, wooden-soled shoes to protect their feet from flying sparks.

In the 1920s, continuous hot rolling of steel was perfected, followed by the development of continuous cold rolling techniques. In the 1930s, an electrolytic method for plating strip steel was developed, and the process of manufacturing tinplate became entirely automatic.

The growing market for preserved foods was the primary impetus for the growth of the tinplate industry during the nineteenth century. Consumer use of canned foods was not substantial until the latter part of the century, however. Initial stimulation of the canning industry was provided by government interests. Both British forces during the Crimean War and American soldiers during the Civil War relied heavily on canned foods, and the military forces of several nations promoted and financed food preservation experiments throughout the century.

Before the nineteenth century little was known about the preservation of food by canning. Fish and meat were salted and sometimes dried, fruits and vegetables were usually dried and many foods were stored in cool, dry rooms for use during the long winter months. Recipes from 1680 have been found for the making of jam, but for the most part, the coming of winter was a worrisome prospect. Even a bountiful harvest could spoil and if crops had been meager, food stored inadequately could result in a critical food shortage.

Soldiers and sailors were particularly affected by the problems of food preservation because they relied on preserved foods. Sailors attempted to take fresh food with them and sometimes carried an edible cargo of live animals, but all too often the animals were injured or became ill in transit, and the sailors subsisted on salted meat and biscuits. Without fresh food, they were threatened constantly by scurvy, a vitamin deficiency disease which could debilitate an entire crew, rendering a ship helpless.

Soldiers were hardly more fortunate. It was customary for armies to sack the surrounding countryside to weaken their enemies. As a result, soldiers were forced to rely on provisions carried with them. Since these supplies rarely included fresh foods, soldiers also were often attacked by scurvy. To make matters worse, the causes of scurvy were not understood. It was thought overindulgence in salted meat provoked the disease, and during the Crimean War, English troops suffering from the disease were given dried fruits and vegetables instead of fresh.

Napoleon, who understood the problem of feeding large armies, offered a prize of twelve thousand francs to anyone who could invent a process for keeping food fresh. In 1809, French chef and confectioner Nicholas Appert was awarded the prize by the *Bureau Consulatif des Arts et Manufactures*. In the following year he published a massive work containing instructions for preserving a wide variety of foods. His work, *Le Livre de tous les Ménages ou l'Art de Conserver pendant Plusieurs Anneés Toutes les Substances Animales et Végétales* was reprinted five times in French and four in foreign languages. For over fifty years it remained the

5. This Maryland Chief can originally held oysters. It was made in the early 1900s, and is an example of an open-top or sanitary can. The same design, practically unchanged, is in use today.

foremost reference on the subject.

Appert's process was simple. Allowing space for expansion, he placed the food to be processed in large-mouthed glass bottles which were stoppered tightly with corks and wire and cooked in boiling water. The cooking time varied according to the kind of food to be preserved. Appert did not realize that microorganisms are present within the foods themselves. (Pasteur did not complete his research until 1860.) He believed his methods were successful because of the total exclusion of air brought about by the heating.

Despite Appert's theoretical misconception, his detailed instructions and results are beyond reproach. He stressed proper methods and equipment. Suitable wide-mouthed bottles were not available, so he blew his own. His meticulous canning instructions include the precise times for canning a wide variety of meats and vegetables, as well as juices, soups from his own recipes and even milk. He stressed the importance of the freshness of the food to be canned, and recommended that produce be brought straight from the garden to the bottle. His calculations of cooking time made allowance for food picked late in the season and his instructions also included serving suggestions.

The same year Appert published his findings, both Augustus de Heine and Peter Durand obtained English patents for the use of metal containers in the canning of foods. Neither produced canned foods on a commercial basis. But in 1811, Bryan Donkin and John Hall patented a food-canning process derived from Appert's methods. Later, they stated they had purchased a patent from Appert, but there had never been any patent to be sold and no record of such a sale occurs in the Frenchman's files.

Donkin and Hall found glass too fragile and cork stoppers too porous, and converted quickly to soldered tinplate containers. In 1813 and 1814, their canned meats were tested by the British Army and Navy. Reports were favorable, and in 1818, Donkin, Hall and a new partner, Gamble supplied over forty thousand pounds of canned foods to the British Navy. The favorable reports continued. Sailors were delighted with the canned meats and vegetables, and commanders wrote testimonials praising the products. In 1820, Captain Sir Edward Parry returned from his first Arctic expedition with hearty praise for the canned goods he had taken with him. A

THE TIN CAN BOOK

6. Originally can manufacturing was a cottage industry. The local tinsmith made a few cans along with his other tinplate articles. A skilled craftsperson could only fashion five or six cans an hour.

later expedition discovered several cans of food left behind during Parry's third voyage of 1825. The contents were reported to be perfectly preserved. By the 1830s, canned foods began to appear in British shops.

Canned foods were introduced into the United States in two places at the same time. In 1819, Ezra Dagget and Thomas Kensett began packing salmon, lobsters, and oysters in glass in New York. The same year, Charles Underwood began marketing fruits and vegetables in crocks in Boston. When the panic of 1837 raised the price of glass, both canners began to use metal containers.

The heavy iron cans of the early nineteenth century were especially clumsy. First a tinsmith cut tinplate into the desired shapes and then laboriously hand-soldered them together. The can body was shaped first, and the side seam soldered together afterward. Next, the bottom of the can was flanged and soldered onto the body. Finally, a top with a circular opening about one inch in diameter was soldered in place. The food to be canned was chopped small enough to fit through the opening at the top (photograph 7).

After the food and canning liquid were in the can, a small tinplate insert with a pinhole was soldered over the opening (photograph 8). Then the can and its contents were boiled until steam began to escape through the pinhole. A drop of solder was placed over the pinhole, and the cooking continued until the processing was complete. Finally, a ring was attached to the top of the can, and an embossed metal or paper label attached to the front. Sometimes a metal wire or thin metal strip was soldered onto the can with the lid. This strip could be pulled to break the solder and remove the lid.

THE TIN CAN BOOK

The thick, heavy cans and their finely chopped contents are hardly appealing to modern tastes. But nineteenth-century American people were spreading rapidly across the country, and canned goods were a boon to travelers. Gold rush adventurers and homesteaders in covered wagons welcomed finely chopped food in ungainly containers.

The Civil War brought canned goods into the American home to stay. Canned goods were widely used during the conflict, and returning soldiers, familiar with food in cans, attested to their wholesomeness. More than any other canned food they praised evaporated milk produced by one of the century's canning heroes, Gail Borden.

Gail Borden may be the most benevolent figure in canning history. He was a well-traveled man of many interests and occupations. He spent his early life in the midwestern United States where he worked as a farmer, a land surveyor and a teacher before becoming interested in the science of food preservation. In the 1830s, Borden experimented with pemmican, a substance composed of a mixture of dried meat and fruits, for Dr. Elisha Kent Kane's Atlantic voyage. His product was successful. In 1851, he traveled to Europe to receive the

7. Cannery workers pack hole-and-cap cans with finely chopped oysters. A similar can was first patented in 1810. By 1922, the success of the sanitary can had made the hole-and-cap design obsolete.

THE TIN CAN BOOK

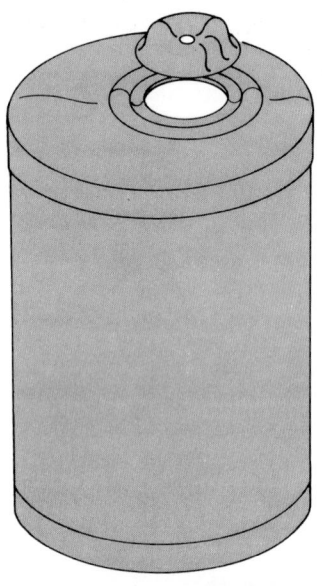

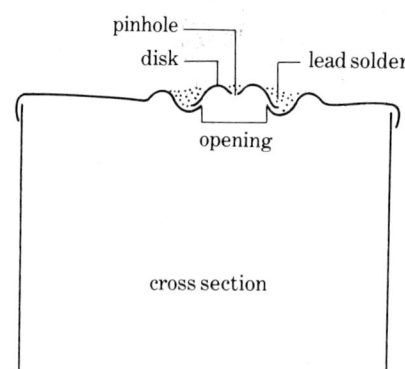

8. After food was put in a hole-and-cap can, a disk with a pinhole was soldered over the opening. The can was cooked until steam escaped. A drop of solder was then placed over the pinhole.

9. Workers in a can manufacturing plant attach tops and bottoms to hole-and-cap cans. Machines capable of crimping can lids over food containers were not invented until the early twentieth century.

THE TIN CAN BOOK

10. Four round-eyed infants encircle a can of Gail Borden's condensed milk. The picture is from a nineteenth century trade card. Manufacturers' agents distributed these cards to shopkeepers.

Great Gold Medal at the Great International Exposition and was elected to the London Society of Arts. Upon returning to the United States, Borden began to experiment with canned milk. He invented a method of evaporating and sugaring milk which allowed it to be canned without damage. At first he was unable to obtain a patent for his process because the patent office insisted his idea was not new. Even Nicholas Appert had proposed methods for condensing milk with the addition of sugar to prevent discoloration. Finally, in 1856, after his third attempt, Borden was issued a patent for the processing of canned milk under vacuum conditions.

At the outbreak of the Civil War, Borden's canned milk business was not going well, but government purchases brought him both wealth and fame. Borden's condensed milk was credited with saving many lives during the conflict. As the product spread to remote regions of the country, it was also credited with significantly lowering the infant mortality rate. Borden's reputation for charity was further enhanced by his donations of fruit juices to Civil War soldiers, and of pemmican to westward-bound travelers. But most importantly, Borden's milk convinced the American householder that canned foods were safe and wholesome.

During the first half of the nineteenth century, the manufacture of cans was a cottage industry (photograph 6). The can maker was often the canner as well. Tin containers were made in the canning plant during the winter and filled in the summer. But as the demand for cans increased, can making became more fully automated and a separate can manufacturing industry developed.

In the early 1800s, a skilled workman could produce only five or six tinplated canisters per hour. In 1847, a pendulum press was patented that was able to cut out can ends from tinplate sheets (photograph 11). Soon afterward, pedal-operated fixtures were invented to hold can bodies in place while they were soldered. By the 1870s, production had increased to sixty cans per hour.

Can manufacturing plants began to spread across the country. One of these firms was opened by Edwin Norton in Toledo, Ohio, in 1865. In 1869, Edwin moved his firm to Chicago, and convinced his brother Oliver to join him there. During the 1870s, Norton Brothers Company manufactured tea and coffee caddies for grocers, but by the 1880s, its emphasis had shifted to the production of vegetable cans. In 1883, the brothers invented a semi-automatic body maker which mechanically soldered side seams. The Nortons' new machinery increased can production to twenty-five hundred cans per hour. Ten years later this astronomical figure more than doubled when they introduced the first fully automatic machine capable of stamping an entire can from a sheet of tinplate.

As production increased, an ever-widening selection of foodstuffs appeared in tin containers. By the end of the century, Joseph Campbell was selling ten ounces of soup for a dime; by 1900, sixty three different kinds of meats were available in cans. In 1909, tuna fish was canned successfully, and in the 1920s citrus fruits and tomato juice appeared in tin containers.

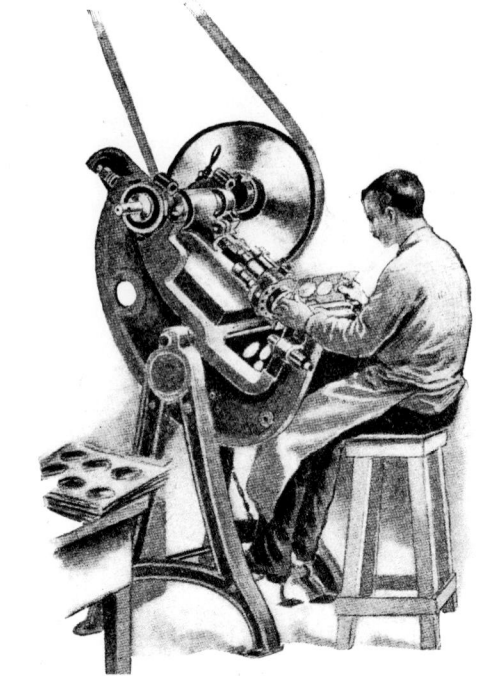

11. A pendulum press for stamping out can ends was patented in 1847. Machinery increased individual worker production from a few cans per hour to fifty or sixty cans per hour.

The first major advancement in twentieth century can making was the formation of Sanitary Can Company in 1904 by New York City jobbers Bogle and Scott, Max Ams Machine Company of New York, and Cobb Preserving Company of New Jersey. This firm first manufactured the sanitary tin can as it exists today. The sanitary or open-top can was constructed during a completely mechanical, double seaming operation that required no soldering: a rubber compound held the lock-and-lap side seams together. The top of the container was completely open. The lid was crimped on automatically after the can was filled with food. Consumers were conservative, and it took time for the sanitary can to be fully accepted. But by the 1920s, the hole-and-cap was a thing of the past.

The hermetically-sealed vegetable can was not the only product enjoying the benefits of the expanding markets and technological advances of the last quarter of the nineteenth century. The makers of metal boxes, sometimes but not always the same firms which manufactured vegetable cans, were also doing a brisk business: " . . . [T]ea and coffee, matches, tobacco, jams and bonbons, soup, fish, arrowroot, revalenta baking powder, blacking and the many other articles in daily use . . . handsomely covered with paper labels, or still more handsomely finished by the new decorating process, are to be seen in shop windows wherever you pass," Philip William Flower wrote of England in 1880 in *A history of the trade in tin*.

Actually, metal boxes predate vegetable tins. Some (like the handpainted tea canister, photograph 14), date from the late eighteenth or early nineteenth centuries. One early metal container was a small lead drum manufactured in the 1780s, as a snuff box. Fifty years later, Huntley and Palmers, English biscuit makers, and John Walker, match maker, began to package their wares in tin boxes. Walker sold matches in tins at a slightly higher price than matches in cardboard boxes, and Huntley and Palmers provided shopkeepers with returnable tin containers.

Both John Walker and Huntley and Palmers chose metal packages for purely practical reasons. In the 1830s, there were no safety matches, and tin packages offered Walker's product significantly better protection against moisture. Huntley and Palmers began packing biscuits in tin containers when their business expanded to a national and then international scale.

Huntley and Palmers' biscuit tins mark the beginning of the era of collectible tin

WILLIAM VOGEL & BROS.

PATENT COVER CAN.
This can is provided with a cover, which being replaced after the top of the can (which is thin tin) has been cut out, will enable the consumer to preserve the contents.

PATENT STRIP CAN.
Removing the STRIP from this CAN leaves the cover loose, ready for further use or for protecting contents from dust, &c.

THIN TOP CAN,
For Paints, Condensed Milk, &c.
The top of this can is made of very light tin, and is easily cut out with a pen knife.

Special Lists, Samples and Prices furnished on application.

12. Food cans of the third quarter of the nineteenth century were particularly cumbersome. Several methods were developed to make opening the heavy iron containers easier; all involved purposely weakened sections of the can.

13. The original Huntley and Palmers tins were plain square boxes made for the biscuit manufacturers by Huntley Boorne and Stevens. This tin is a tiny twentieth century replica of the original nineteenth century tin. This photograph is larger than life: the actual box measures 1½ inches (3.8 cm) by 2 inches (5 cm). Lithographed paper labels were usually glued on the outside of the original 12 inch (30.5 cm) square tin box. The replicas come in several colors; this label is blue.

containers. The biscuit shop, originally operated by Joseph Huntley in Reading, England, was lucky enough to be located across the street from the inn where the coach from London to Bath stopped. English inns of the time were notorious for serving food of poor quality at high prices. Many passengers carried snacks of their own, and were more than willing to run across the street to buy a few tea biscuits from Huntley and Palmers. Soon the bakers saved them the trouble and sent a boy to the inn with a basket of biscuits for sale. Since the biscuits were good-tasting and well-made, many travelers requested them from their local grocers when they returned home.

Because their biscuits were fragile and spoiled easily from moisture, Huntley and Palmers needed to ship them in air-tight, durable containers. The bakers solved the problem by commissioning metal boxes from Huntley Boorne and Stevens, ironmongers conveniently located a few doors away.

The first tins made by Huntley Boorne and Stevens were plain, square tin boxes. Huntley and Palmers affixed a paper label (photograph 13) and sent them to grocers across the country. These early bins were designed for use in stores. The grocer sold biscuits from a Huntley and Palmers tin on his counter. The tins not only displayed the biscuits attractively, they also kept the biscuits fresh. When the tin was empty, the storekeeper returned it to the bakery.

14. This heavy, tinplated, iron tea canister could have been made as early as 1790, and certainly dates from the early nineteenth century. The label was hand painted on the tin. The scene, showing a lion and hunter, indicates that the tea was of African origin. The numeral "4" above the label refers to the type of blend. The tin is large: 17 inches (43 cm) tall and 9 inches (23 cm) in diameter.

THE TIN CAN BOOK

15. Printed paper labels were available early in the nineteenth century, but they were easily lost. Before the development of processes for printing on tin, manufacturers soldered embossed labels onto their cans.

16. The entire lid of this early pocket tobacco tin is embossed. Embossing techniques performed two functions. The label was permanent, and the tin was decorated.

17. Stenciling was also used to label tin cans. It was faster than handpainting and did not require metalworking tools as did embossing. These two coffee cans were made in the 1840s or 1850s.

However, paper labels were not entirely satisfactory. The spread of tin boxes meant that a manufacturer could sell his product nationally, but the size of the market meant that the manufacturer could not be present when the product was sold. Paper labels were easily lost, and an unscrupulous seller could replace the correct label with one of his own.

The first method of making a permanent label was simply to paint a picture on the tin. The tea canister (photograph 14) is one example of a very early handpainted tin. It assumes no literacy among its customers. The lion and hunter probably indicate that the tea inside is of African origin; the number is a reference to a particular blend. Handpainting made the label permanent but it was extremely time-consuming.

Early can manufacturers soldered em-

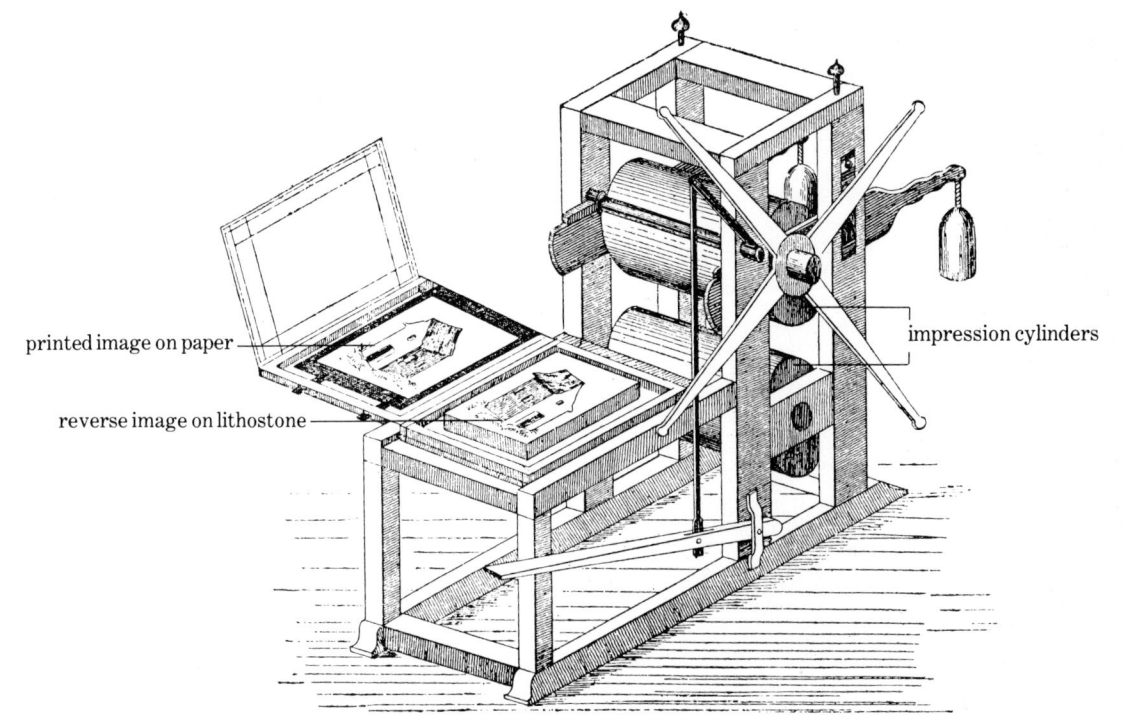

18. Senefelder's lithographic press was hand operated. The paper to be printed was placed on top of the inked lithostone, covered with a tympan sheet and rolled through the press.

bossed labels to the can (photograph 15). Embossing the can itself was even better insurance against losing a label (photograph 16).

By the middle of the nineteenth century, metal box makers were experimenting with several different kinds of labels in addition to paper, embossed and handpainted designs. As early as the 1830s, Huntley Boorne and Stevens decorated tins for Huntley and Palmers using a *moire metallique* finish. The tinplate was subjected to very high heat; the result was a crackled effect involving crystallization of the metal surface. Stencils were also used (photograph 17). Tins to be stenciled were painted a solid color and dried. An image or name was applied in a second color.

Transfer printing had been used successfully on pottery jars a century earlier, and the technique was suitable for tin boxes. When an image is transferred, it is printed in reverse on transfer paper and then pressed onto the desired surface. Very complicated decorations, after being lithographed on transfer paper, could be transferred onto metal surfaces. Paper labels were also lithographed.

Aloys Senefelder invented the lithographic process in Bavaria in 1798. He was an actor and playwright who was having difficulty with his local printer. As a result, he began experimenting with printing methods of his own. By accident he discovered that a drawing made on limestone, using a greasy pencil or crayon, could be inked and reproduced. Senefelder worked out the details of his process and patented the invention in the early part of the nineteenth century. The technique was not ex-

19. This collar box from the 1870s was designed to have a secondary use as a tobacco tin. It is very important because it was printed using a direct process involving roughening the surface of the tinplate.

traordinarily difficult and its reproductive qualities were excellent. Soon lithographic shops were opened throughout Europe.

The principle involved in lithography is simple: grease and water do not mix. Senefelder discovered that by treating different sections on the surface of a limestone block, it is possible to make some sections water attractive and ink repellent. Other sections become ink attractive and water repellent. With a process involving inking the stone and damping it with water, a properly treated limestone lithostone can be handled like any other printing surface. An image can be reproduced hundreds of times.

Lithography is also called planography. This technique is particularly interesting because it does not involve any true etching of the stone. Unlike a truly etched surface, on which the surface of the printing plate is chemically dissolved away from the elements to be printed, the limestone surface remains on the same plane. It is physiochemically treated in a way that gives different sections different properties.

The process is not difficult but requires several steps. First the lithostone is sensitized so grease adheres readily to it. Then the image is drawn on the stone with a greasy crayon or pencil. The drawing may be accomplished with as much tonal variation as any drawing; the sensitized stone reacts readily to thin or thick crayons and differences in pressure. After the drawing has been made, the entire stone is treated. The grease is caused to combine chemically with the stone, and the remaining surface is etched away slightly from the greasy areas. This etching is not deep enough to change the plane of the stone. Then the stone is washed and inked, and the etching process is repeated. The greasy areas are not sufficiently ink-receptive to allow the image to be printed repeatedly after the first etching. But it is possible to ink the stone and to print a sample proof at this time. After the stone has been etched a second time, it is possible to print repeatedly without damaging the printing surface.

A lithograph may be printed in one color or in several, but a separate plate or stone is required for each color. Color printing may be accomplished in distinct areas or several colors may be printed on top of each other. Printing colors on top of each other requires skill in calculating the resulting combinations of tones and intensities. Considerable depth and subtlety can be obtained by using several stones inked with different colors. Both methods require great care in the printing. Each new stone must be perfectly aligned with any previous printings. If the colors are even slightly out of line, the print will be ruined.

During his initial experiments, Aloys Senefelder developed thin paper on which an image could be drawn and then transferred onto a lithographic stone. (One problem in lithography is that the image must be drawn on the stone in reverse.) This paper made it possible to print on metal using indirect lithographic techniques. Using a lithographic stone, the image was printed first in reverse on transfer paper. The transfer was then applied to a flat sheet of tinplate. With this method it was possible to print several colors at once. The transfer paper was printed in reverse order, with the last color printed first and the first color last. Then the entire image was transferred under great pressure onto the metal.

Since transfer printing involved a time-consuming hand process and had little real potential for multiple reproductions, numerous attempts were made to adapt the lithographic process to direct printing on metal. But it took time to develop a suitable

20. Ginna and Company of New York was one of the finest late nineteenth century lithographers. This Rip Van Winkle story tin was printed with twelve separate stones in different colors. It was made for Holmes and Coutts, New York biscuit makers.

EVALUATING TIN CONTAINERS

A discussion of the condition or rarity of a tin can assumes that the container being evaluated has been skillfully made, and labeled with interesting and high-quality graphics.

NEW OR MINT CONDITION

Tins in this category show virtually no evidence of handling. They are in perfect condition. Even close scrutiny reveals no surface flaws.

ALMOST NEW OR ALMOST MINT CONDITION

These containers are near-perfect. The colors are clear and bright with no evidence of fading. Upon close examination, however, small defects are apparent: slight traces of rust, nearly imperceptible scratches, minor dents or slight chipping.

GOOD CONDITION

Tins in the middle category may show some fading of colors, though all printing and detail work should be clear. Also, there may be some easily discernible surface imperfections: scratches, small dents or corrosion. These defects, however, should not interfere with the overall appreciation of the tin.

FAIR CONDITION

Containers in fair condition are still worthwhile examples, but somewhat lacking in general appearance. The surface is not severely damaged, and large portions of the graphics are not obscured. But the tin may be rough, faded or show noticeable corrosion and blemishes.

POOR CONDITION

Tins with severe defects, heavy corrosion and extreme fading are of little value and should not be considered except in cases of extreme rarity or very low price (free, for example).

It is almost impossible to fix a price on a collectible tin container. Prices fluctuate enormously over extremely short periods of time. One collector might pay many times what another would for the same item. For instance, if a tin is needed to complete a set, it may be extremely valuable to one collector but have no special value to another. If a tin is the only known example of a particular design it may be generally considered very valuable. But when a second example is discovered, both will be devalued.

    A beginning collector can become knowledgeable about pricing tin containers only through experience. Of course, mistakes will be made. The best advice for beginning collectors is to buy things they like. In this way value to them will be assured.

    A tin container in good condition is collectible. If it is also rare, it may be valuable. The dollar value of any tin container is determined by a straightforward economic rule: the amount a seller will accept, and a buyer will spend.

method because both metal and stone are hard substances. If ink is pressed between two hard surfaces, the image becomes "smashed"—smeared and indistinct. An additional problem was the lack of inks suitable for printing on metal. Unlike paper, metal is not porous, and the ink must adhere to the surface. Available inks did not adhere properly. Attempts were made to roughen the metal surface in order to make it more receptive to ink, but to little avail. To make matters worse, the pressure created by printing often broke the lithostone. Single-color lithography was not successfully applied to metal until the 1870s, and color lithography (chromolithography) was not used until the 1890s.

Mid-century printers were conservative. In spite of the success of lithography and chromolithography, they still preferred more old-fashioned thin-line engraved designs. Color was introduced cautiously, first by printing or transferring an outline in black, and then adding color by hand. Currier and Ives, for example, used this method to color their prints. The excitement of French poster art during the last thirty years of the nineteenth century was responsible for at last turning the attention of printers and fine artists to the enormous potential of color lithography. Subsequent developments in printing machinery eventually produced satisfactory tin-printing equipment.

In the meantime, tin boxes, especially those made from many separate pieces of tinplate, were often colored with combinations of techniques. The outstanding pagoda-shaped store bin (photographs 29-31) was manufactured in the 1870s, and embodies several different printing methods. First the entire surface was covered with a base coat. Next, lettering and stylized designs were stenciled onto it. More complicated artwork was probably transferred, and final touches were added by hand. Even twenty years later, when color lithography was available, this kind of tin would have been colored almost the same way.

Events during the 1860s foretell the coming breakthrough of lithographic printing on metal which was first perfected in England. In 1862, a Tin Plate Decorating Company was established in South Wales. In 1864, a patent was issued for lead-based inks suitable for printing on tin, followed in 1869 by another for the use of a paint or alloy on tinplate and in 1870 by a third for the use of clear lacquers. Also in 1870, the London printing firm of Barclay and Fry invented the offset printing press. Using a flatbed press, Barclay and Fry printed a lithographic image onto a rubber blanket wrapped around a roller. Next the image was printed (or transferred) onto paper or metal fed between the rubber roller and a second roller. The press carried a lithographic stone, which moved horizontally back and forth with the two rollers mounted above. After the rubber roller picked up the image from the lithostone, it was rolled onto the moving paper or tinplate. By printing from hard-to-soft-to-hard the problem of metal lithography had been solved.

The earliest lithographed tin containers were printed in one color (photograph 19). First a solid base coat was applied to the tinplate and dried. Initially, the base coat was air-dried, which could take an entire day; later, it was heat-dried or "stoved" in a furnace. When the base coat was dry, the design was printed in a second color and dried again. Sometimes a final coat of varnish was added and the tinplate dried a third time. The decorated tinplate was shaped into a container after the printing was completed. Sophisticated stamping machines made it possible to form containers without damaging the artwork.

Color printing became practical twenty years after the first tins were lithographed, but it was a long, complicated process. Chromolithography is complicated enough for paper, but metal-printing required several additional steps. For example, it was discovered that printing a coat of varnish between coats of colored ink not only enriched the final shades but altered slightly the inks underneath. Sometimes several coats of varnish were printed between ink printings. Also, two printings of key colors, especially white and red, were necessary to guarantee the desired intensity. Sometimes the tinplate was run through the press, dried, and run through again using the same plate and color. Occasionally, a second plate with minor variations was printed in the same color to create further tonal differences. And after each color was printed or treated with varnish, the metal had to be dried. It sometimes took several days to complete one tin container design. Ginna and Company, famous for the beauty of its colored tins (photograph 20) once produced a container requiring twelve different plates.

The last third of the nineteenth century not only witnessed the successful adaptation of lithographic printing techniques to metal, but also the formation of major American tin container manufacturing companies. In 1865, Colonel Silas Augustine Ilsley began making tinware in Brooklyn, New York. His firm expanded quickly from twenty to two hundred employees. Other tin container makers had similar good fortune. In 1869, Guy A., Daniel M., and Joseph L. Somers began making cans by hand in one room in Brooklyn. By 1883, their factory occupied eight rooms and employed between one hundred and fifty and two hundred employees. Born inventors and mechanics, the three brothers not only manufactured tin con-

21. Norton Brothers specialized in store bins which could be sold to shopkeepers in sets. The revolving label featured a patented roller which turned to display the names of eight kinds of tea (or coffee).

THE TIN CAN BOOK

22. Ginna and Company was as expert at one-color lithography as chromolithography (see photograph 20). These beautiful Santa Claus tins (the red tin on the left is shown from the front and the green tin on the right from the back) show the wide range of tonal differences and intricacies of design possible in high quality tin printing. Both tins were made in the 1880s, and are 8 inches (20 cm) tall.

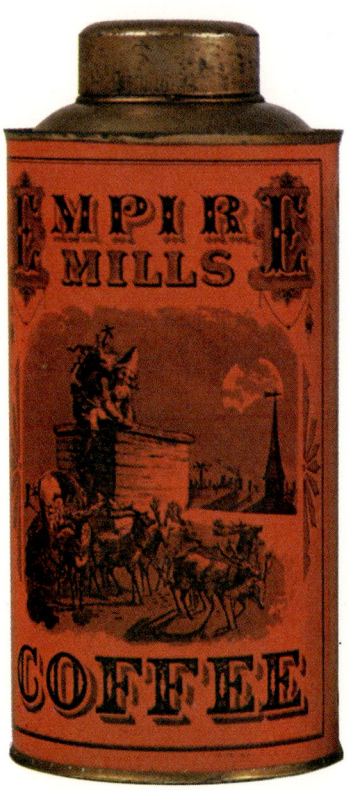
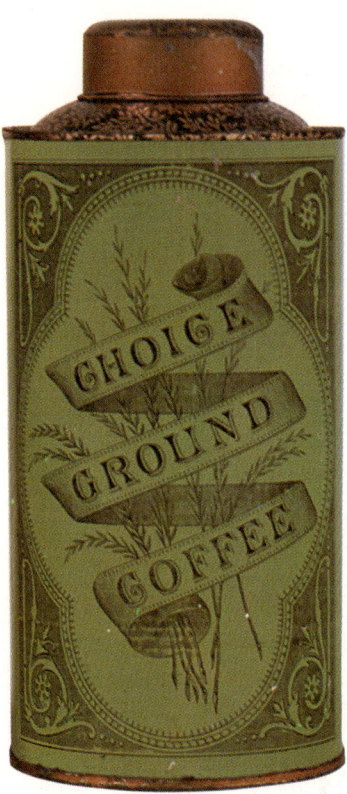

tainers but designed and built their own dies and stamps. In 1879, they patented a process for printing on metal.

In 1875, the Norton brothers had already expanded their facilities in Illinois a second time, and Charles Hazlewood Hasker was making small paper boxes and tobacco tags in a woodshed behind his house in Richmond, Virginia. In 1891, Hasker joined with grocer Milton Marcuse and Marcuse's father and brothers to form Hasker and Marcuse Manufacturing Company. At first the firm manufactured lithographed metal tobacco tags and then began producing tin tobacco containers. With one hundred tobacco manufacturers in the city of Richmond alone, they were in an advantageous location.

Each of the early firms had its specialty, and their workmanship is recognizable. Ginna and Company, and Somers Brothers made detailed tins with fine line drawings against a solid color background. But the excellence of the chromolithography on a Ginna tin (photograph 20) is unsurpassed. The Nortons, who produced mainly vegetable cans and can-making machinery, also specialized in store bins. Their complicated, elaborately decorated bins show a marked Oriental influence (photograph 21).

On the other hand, by the time Hasker and Marcuse began making tin containers in 1891, color lithography was the rule, and their tobacco tins are nearly all in color. Somers Brothers was responsible for some early, ingenious containers. For example,

23. Independent firms of the late nineteenth century marked their containers. The presence of some of these marks means a tin was made before 1901, when the industry was consolidated by American Can Company.

the firm created one of the early Mennen talcum powder tins, featuring a rotating top which moved to alternately cover and uncover dispensing holes.

At the same time, the tinplated canister industry was also flourishing in Great Britain. Huntley Boorne and Stevens were licensed to use Barclay and Fry's offset press for printing tinplate in the 1870s, and began manufacturing lithographed tin boxes for Huntley and Palmers. Occasionally, but not often, the firm made boxes for other biscuit manufacturers. Several dozen English candy and biscuit makers packaged their products in tin containers, and these English biscuit tins are among the most elaborate tin containers in existence.

Almost from the start, the emphasis of British manufacturers was different from the American. The British recognized the advantages of printing directly on metal containers as quickly as the Americans, but instead of stressing an identifiable trademark or brand name, they often concentrated on secondary uses. They reasoned that a beautiful, reusable container kept their product in the consumer's mind. A tradition of packaging in deluxe containers for special occasions had existed for some time before the development of tin boxes.

The numerous Christmas tins of the biscuit makers show the British secondary-use theory pushed to its limit. The gift items were principally manufactured for Huntley and Palmers by Huntley Boorne and Stevens, although other manufacturers offered competitive designs each year. The firms made a range of extravagantly shaped and intricately colored tins. Some tins hardly looked like biscuit containers. They were made to resemble stacks of Staffordshire plates, concertinas, perambulators, china cabinets, fishing creels and houseboats. The manufacturer's name was generally on the bottom of the tin or found on the inside cover, which is a custom retained to the present day. Less pretentious tins of more ordinary shape were produced for every day occasions.

Occasionally, American metal box manufacturers also produced tins with a secondary use. The most notable examples are the tobacco lunch pails of the early twentieth century. But manufacturers mainly preferred graphic designs which stressed the brand name of the product. If the brand name was "Tiger," the picture of a tiger was placed on the can. Later, eye-appeal was given a high premium, and the trademark was stressed almost at the expense of the product. Nineteenth century American containers, though not necessarily reusuable, sometimes cost more to produce than their contents.

The era of the small, independent tin container manufacturer in the United States collapsed only three decades after it had begun. The turn of the century ushered in the era of the great trusts, and the tin can industry was one of many industries caught in extensive consolidation. William H. Moore, Daniel G. Reid and William B. Leeds, the same three men who had organized American Tinplate Company in 1898, next turned their attention to individual can companies. In 1901, they created American Can Company by effecting a merger of one hundred and twenty three factories, representing sixty different firms. Edwin Norton became president of the new organization, and the Norton Brothers' plant in Maywood, Illinois became the center of the conglomerate's operations. The simultaneous demise of so many small firms accounts for one method of dating tin containers. For example, if the can bears the name of an independent firm like Ginna and Company, or S.A. Ilsley, the can was made before 1901.

Individual can manufacturers reacted differently to the formation of American Can. By 1901, Hasker had already sold his interest to the Marcuse family and retired. After American Can's takeover, S.A. Ilsley retired to the Green Mountains of Vermont. He raised Morgan horses in the town of Middlebury, built a church and library, and became known as a benefactor. The Somers Brothers' plant became one of the major factories in the American Can network. Edwin Norton remained president of American Can until 1904, when he left to form Continental Can Company.

Although American Can made a determined effort to monopolize the can-making industry, its impact was strong but brief. Many of the factories purchased by the trust closed after only a short time. Continental Can offered some competition almost from the start, and National and Pacific Can also captured a small but significant share of the market. In 1916, the passage of anti-trust legislation severely restricted American Can's activities.

While some can companies escaped the merger, others continued to be formed. Tindeco (Tin Decorating Company; Baltimore, Maryland) began manufacturing tin containers in 1900 and became a major can producer, and Heekin Can Company opened its doors in 1901. Many small firms resisted the pressure of the trusts, and some turn of the century tins bear the legend, "Not made by a trust."

During the early twentieth century the canning industry continued to change. The Food and Drug Administration, passing its first act in 1906, limited the claims a manufacturer could make on a product label. Canned foods were consumed in steadily increasing amounts, and canners found it economical to become directly involved in the growing of vegetables.

Advertising was also changing. The retailer no longer relied on point-of-purchase marketing. Advertisements in national magazines, on radio, and later on television sold products long before consumers saw them in stores.

Advancements in printing were occurring on a regular basis. In 1895, zinc plates were introduced as a replacement for the breakable, cumbersome lithostones used in flatbed presses. (Senefelder had experimented with metal lithographic plates as early as 1818.) The lightweight metal plates were particularly well-suited for rotary offset presses (photograph 24). The first rotary offset press was installed at the Somers plant, now a part of American Can, in 1908. Two-color rotary offset presses were also developed, but two one-color presses were usually used to print two colors.

THE TIN CAN BOOK

In 1914, the introduction of continuous ovens for drying inked tinplate made it possible to decorate metal sheets in one operation. Now tinplate could be printed, stoved and printed again without being handled. During the 1930s, a four-color printing process was developed in which all colors were broken down into four basic colors. Any color could be printed with a combination of the four basics. Photolithographic techniques were also improved. Tin container collectors agree that by this time originality in tin container labeling and design was on the decline. English biscuit tins, however, continued to be imaginatively made until the 1940s.

But the era of the collectible beer can just begins in the 1930s. Beer was first bottled in England in 1866; in 1909, a brewer in Montana made the first attempt to can beer. Unfortunately, his experiments were doomed to failure. Beer must be canned under pressure, and the cans of the day were not strong enough. The side seams were strained and the ends buckled. Furthermore, the tin coating inside the can combined chemically with the beer to create an unsightly and foul-smelling brew.

Experiments with enamel linings for cans had been conducted since 1868. In the 1920s American Can perfected a compound called "C-enamel" to line cans holding highly corrosive substances like citrus fruits and fruit juices. But even C-enamel was not strong enough for the corrosive action of beer. The enamel coating dissolved, turning the beer milky.

Attempts were made to coat beer cans with pitch, which was used to coat the inside of wooden beer kegs, but pitch was difficult to apply and unattractive in appearance. Finally, American Can developed a two-coat process in which the cans

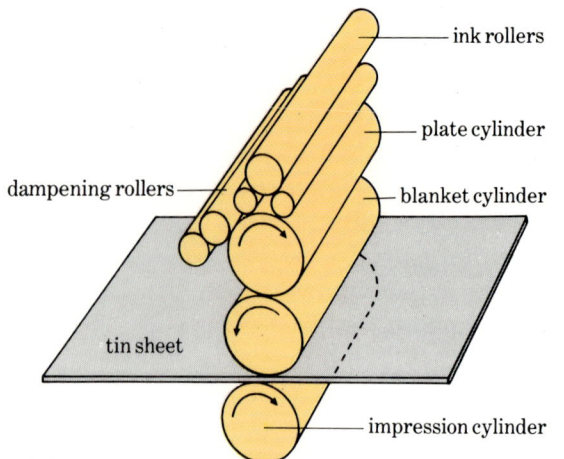

24. The era of the decorated tin container began with the introduction of the offset press in 1870. A soft rubber blanket cylinder carried the image from the hard printing surface to the hard tinplate. Early offset presses were flatbed. Rotary presses like this one were not used until 1906.

were coated with a tough, impermeable chemical lining and then sprayed with a second substance.

In 1935, the first commercial flat-top beer can filled with Krueger Cream Ale was test-marketed. The beer sold surprisingly well, and soon Pabst Brewery began to distribute its Export label in American Can's flat-tops.

Later the same year, Continental Can introduced its version of the beer can, the first cone-top beer can (photograph 25). Although Schlitz, a major brewer, was the first beer packaged in a cone-top can, the design appealed primarily to smaller brewers. The cone-top was created in an attempt to utilize the standard bottle-filling equipment of the time; therefore it was more attractive to breweries that could not afford the expensive equipment required by the new flat-top cans. However, the flat-top style was ultimately victorious. They were easier to make, fill and store. By the middle of the 1950s, the cone-top beer can was obsolete.

25. This Schlitz beer can was the first cone-top ever produced. It was made by Continental Can Company in 1935. American Can's flat-top beer can predates it by several months.

26. Iron City Beer was the first to be packaged in a flat-top can with a pop-top lid. The can was made in 1962; the top designed by Alcoa Aluminum.

Early flat-top cans were made in three pieces. Sheets of can bodies were printed, cut out, shaped and seamed, and the bottoms attached in much the same way as with the sanitary can. By the late 1950s, aluminum "soft-tops" were added. They were marketed as a convenience to the consumer, but their reduced weight was a convenience to the brewer and the shipper as well. By the early 1960s, aluminum tops featured the pull-tabs that now dominate the market (photograph 26). The first tabs snapped off, but since 1975 new tabs remain attached to the can top after it has been opened.

In 1959, Coors Brewery produced the first beer can made entirely from aluminum, and by the early 1960s, Reynolds and Alcoa were making all-aluminum cans of their own. A single slug of aluminum is impact-extruded to form an entire one-piece can body. Only the lid needs to be attached. Aluminum cans offer other advantages: they are extremely lightweight and print clearly.

In response to this major challenge, the steel industry followed with a tin-free-steel can made with a thin solderless seam. Tin-free-steel is lighter than tinplate because it is coated with a synthetic compound instead of tin. The less bulky side seam made the lighter cans competitive with the seamless aluminum cans. Soon a process was developed for extruding tin-free-steel and both American and Continental Can began producing two-piece tin-free-steel cans. The new cans have only one major drawback: they cannot be printed before they are formed. Rolling the label onto the curved surface of the can sometimes causes the edges of the design to blur. But experience is rapidly improving this minor setback.

In the 1960s a great controversy in the art world erupted when Andy Warhol im-

27. Andy Warhol's silk-screen of a Campbell's soup can became instantly famous. He was not only criticized for the cartoon-like style but for the subject matter.

28. Jasper Johns made this bronze of a pair of Ballantine ale cans in 1960. His sculpture is dated by the design of the cans: they were made with a three-piece construction and do not have pop-tops.

mortalized a Campbell's Soup can (photograph 27) in his now nearly cliched silk screen. Jasper Johns captured a pair of Ballantine ale cans in bronze (photograph 28) and homages to flashlights, paint brushes and other homely objects ensued. Art critics called the new works "popular" and refused to consider them fine art. The artists, in turn, were affronted by the term "pop" and sneered at the critics.

Now Pop Art has generally been accepted. Modern critics reason it is a response to the present environment and therefore as valid as any other artistic expression. But perhaps criticism against pop objects should be leveled at the society which created the tin can, not the artist who represents it, for the tin can is here to stay.

In pursuing their hobby, collectors preserve the day-by-day objects of the past. In a way, tin container collectors are the historians of a by-gone tin-can age, while beer can collectors are modern-day chroniclers.

In gathering their contemporary treasures, beer can collectors have made their own peace with our throw-away culture. But whatever the reasons, it is obvious that both hobbies bring a great many people pleasure.

There are national organizations for both collectors. The TCCA (Tin Container Collectors Association, P.O. Box 4555, Denver, Colorado, 80204) publishes *Tin Type*, a monthly newsletter available by subscription. The BCCA (Beer Can Collectors of America, 7500 Devonshire, St. Louis, Missouri, 63119) has local chapters in many states. Both associations hold annual conventions—or "CANventions," as the collectors say.

Collectors of tin containers and beer cans are often drawn into their hobbies simply because they find decorated tin containers attractive. Specialization usually occurs after a general interest has been sparked, and then as many different tracks are taken

as there are collectors. Tin containers are classified under the general heading "advertising antiques" and many people are fascinated by the advertising copy of the tins. Other collectors may concentrate on tins made by a particular firm, Ginna and Company, for instance. Still others may prefer tins made before 1900, or specialize in tins of unusual construction. Most often tins are collected based on the product originally marketed in the container. People pursue tobacco containers, pharmaceutical tins, or coffee and tea canisters. Of course, the various categories overlap, and one collector may have several specialties. Almost anyone finds an interesting shape pleasing, a complicated design exciting, or a quaint product name amusing.

The remaining chapters of this book have been divided into some of the general areas in which a collector might specialize. Each begins with a short piece of introductory text; the rest of the chapter is composed of photographs and captions.

Tin containers may be found in flea markets, at auctions, in antique shops and attics. Beer cans are found in dumps, at the side of the road, stashed away in defunct breweries and at the local liquor store. Decorated metal boxes may be found wherever country store collectibles are found. Country stores are primarily a rural phenomenon, and related collectibles tend to be available in such areas. However, the growing interest in tin containers as advertising antiques has shifted their concentration into the urban areas as well, particularly those on the East coast. Beer cans are pervasive. There are dozens of breweries in New York and Pennsylvania, and Illinois, Wisconsin, and California also boast substantial numbers. Large and small breweries exist in nearly all the United States.

SELECTED BIBLIOGRAPHY

Antreasian, Garo Z., with Adams, Clinton. *The Tamarind Book of Lithography: Art and Techniques*. Harry N. Abrams, Inc.: New York, 1971.

Appert, M. *The Art of Preserving all kinds of Animal and Vegetable Substances for Several Years*. Black, Parry and Kingsbury: London, 1812.

Bitting, A.W., M.D. *Appertizing or the Art of Canning: Its History and Development*. The Trade Pressroom: San Francisco, California, 1937.

Bragdon, Charles R. *Metal Decorating from Start to Finishes*. The Bond Wheelwright Company: Freeport, Maine, 1961.

Corley, T.A.B. *Quaker Enterprise in Biscuits: Huntley and Palmers of Reading 1822-1972*. Hutchinson of London: 1972.

Davis, Alec. *Package and Print: The Development of Container and Label Design*. Faber and Faber: London, 1967.

Flower, Philip William. *A history of the trade in tin*. G. Bell and Sons, London: 1880.

Hedges, Ernest S. *Tin in Social and Economic History*. St. Martin's Press: New York, 1964.

International Tin Research and Development Council. *Historic Tinned Foods*. Second Edition. Middlesex, England.

Martells, Jack. *The Beer Can Collector's Bible*. Great Lakes Living Press, Publishers: Matteson, Illinois, 1976.

May, Earl Chapin. *The Canning Clan: A Pageant of Pioneering Americans*. The MacMillan Company: New York, 1937.

Minchinton, W.E. *The British Tinplate Industry: A History*. Oxford at the Clarendon Press: 1957.

Russell, John, and Gablik, Suzy. *pop art redefined*. Frederick A. Praeger: New York, 1969.

Stigler, George J. *The Theory of Price*. Third Edition. The MacMillan Company: New York, 1966.

The Beer Can Collectors of America. *The Beer Can: a complete guide to beer can collecting*. Edited by Larry Wright #2. Great Lakes Living Press, Publishers: Matteson, Illinois, 1976.

The Canned and Packaged Foods Bureau. *A History of Canned Food*. Designed and produced by The Metal Box Company Limited: London, England.

STORE BINS

Large store bins were common articles in the country store. They were storage bins designed for shelf or floor use, and some were enormous, standing several feet high and capable of holding over one hundred pounds of merchandise. All kinds of items were kept in these tins—tea and coffee, spices, cookies and crackers. The purchaser requested the desired amount; the grocer scooped it out and weighed it.

Store bins are some of the oldest collectible tin containers. They were in use several decades before decorated metal boxes were common as individual packages. Tinsmiths supplied the large containers to both wholesalers and retailers of food. They belonged to the food producer and were supplied to the grocer with the product inside, or the grocer purchased them himself and filled them with brands of his choice. In either case, eye-appeal was most important. The bins acted as point-of-purchase advertisements for the food producers. For the grocer, the colorful containers decorated the shelves and stimulated trade.

Although store bins were once quite reasonable, today they are among the highest priced tin containers. Store bins in good condition can sell for hundreds or even thousands of dollars. They are expensive because they are rare: few of the bins were preserved when country stores were dismantled to make way for supermarkets and chain stores. Design patent drawings and catalogues of the era show many large store bins that have not survived.

29
30

THE TIN CAN BOOK

29-31. A treasure hunt that began in the United States patent office yielded this spectacular, pagoda-shaped spice bin. Its present owner found a patent drawing of the tin and in his enthusiasm mentioned it to a non-collector friend. The friend thought he had seen such a tin in the window of an antique shop. The collector rushed to the store and retrieved his prize. The collector is rightfully proud of his discovery. The existence of a patent would indicate that several of the tins were produced, and this is one (of three) known to have survived. It is difficult to affix a value to such an item. The collector can ask whatever price he wishes because his tin is rare. And the tin's value to its owner is extremely high because it is unlikely he will get another.

The bin has six drawers for different spices and is 41 inches (104.1 cm) tall. It would have been purchased by a grocer and filled with brands of his own choice. It was manufactured by Norton Brothers in the 1880s.

The beauty of the spice bin also increases its value. The shape is complicated and involves many separate pieces of tinplate; the decorations are elaborate and necessitated a combination of labeling techniques. Complicated designs like the six central panels (photographs 29 and 30) were transfer-printed, while the simpler single-color designs and lettering were stenciled.

The cast iron figure of a schoolgirl carrying a music book tops this pagoda. A brass ball decorates one of the remaining two examples of the bin, and a cast iron figure of the goddess Diana the other.

31

32

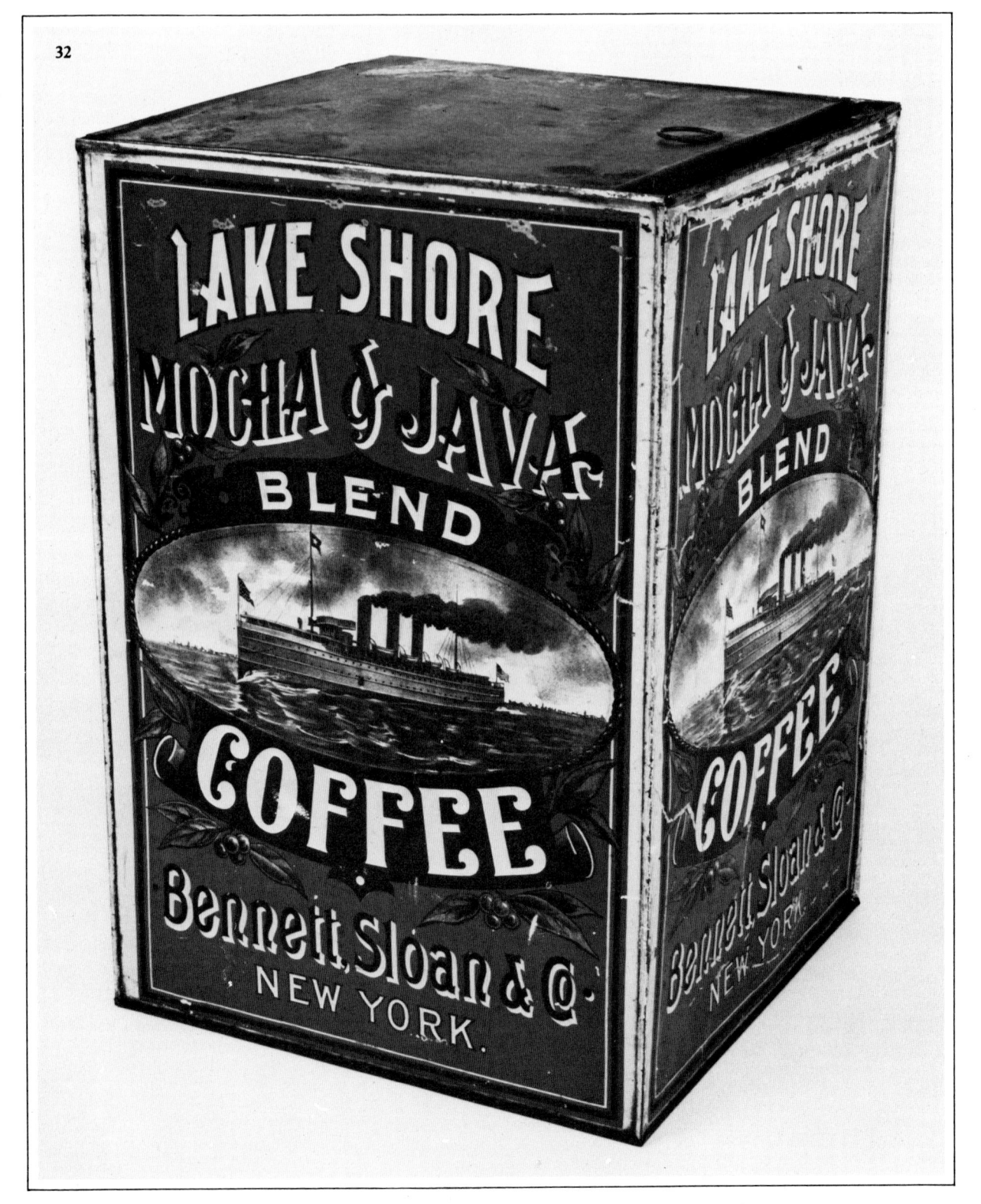

32. This big Lake Shore coffee bin is 20 inches (50.8 cm) tall, and originally held fifty pounds of coffee. This is the only known bin with a Lake Shore label, and the rarity of the tin plus its excellent condition make it valuable. This bin was made by Ginna and Company in the 1890s.

33, 34. This Ginna and Company tin dispensed Irish thread in three weights. The decorations are the blue flowers and green branching stems of the flax plant. They are unusual because they resemble fabric designs, in keeping with the original contents of the tin, and because they advertise the product in an extremely subtle manner. The tin is in excellent condition. Because of the interesting design and the Ginna tin's unusual contents (spice tins occur much more frequently than thread tins), it is highly prized. The box is 5 inches (12.7 cm) tall and the top measures 15 inches (38.1 cm) by 10 inches (25.4 cm). It was made in the 1880s.

35. Ginna and Company was famous for its color lithography, and the lid of this Pastime tobacco-store bin shows why. The hunting scene is not only composed of several colors, each printed consecutively, but the design is also quite detailed and well-drawn. Because of the quality of the image and its size (the lid measures 10 inches (25.4 cm) by 15 inches (38.1 cm), the tin is highly prized as an example of Ginna's expertise. Because the bin was used as a counter display, only the inside of the lid is in color; the exterior is printed with the ornate graphics which also make Ginna tins the favorite of many collectors. It held eighteen pounds of tobacco and was made in the 1890s.

THE TIN CAN BOOK

36. This Plow Boy tobacco dispenser held forty-eight five-cent packages of tobacco—a weight of five pounds. It was made during the early part of the twentieth century. The differences in style and technique between this tin and the other twentieth-century tins on these two pages and the earlier tins on the preceding pages are readily apparent. The tins made by Ginna and Company and Norton Brothers are "old-fashioned." The fine-line graphics on Ginna tins recall nineteenth-century engravings, and the color lithography on both Norton and Ginna tins resembles paintings of the era. The designs on the Plow Boy bin and the others on these pages, however, resemble modern illustrations. The tin is 8.5 inches (21.6 cm) tall in front, 10.5 inches (26.7 cm) in back, and 8 inches (20.3 cm) wide.

37, 39. The Teenie Weenies, little people only a few inches tall, were popular cartoon characters throughout the second quarter of the twentieth century. On the back of this enormous fifty-five pound tin of Monarch peanut butter several families are pictured hefting slices of bread and wielding table knives. The tin was made in the 1920s and is in near-perfect condition. It is 14.5 inches (36.8 cm) tall and 12.5 inches (31.8 cm) in diameter.

38. Log cabins are a particularly popular American motif and appear on all kinds of container labels. This giant log-cabin coffee bin is quite detailed, even showing the interior of the cabin. The calendar on the wall visible through the open door is dated 1913. Through the window two cups of coffee can be seen on the table. The tin held fifty pounds of coffee and is 28 inches (71.1 cm) tall.

39

THE TIN CAN BOOK

40. Rectangular coffee store bins, like this Austin, Nichols and Company bin and the identically-sized Lake Shore bin in photograph 32 were standard in country stores during the late nineteenth and early twentieth centuries. The bins were provided to the grocer full of coffee, simultaneously solving the problems of shipping, storage, and advertising. Such simple prelabeled bins are quite unlike the elaborate store bins made by Norton Brothers to be sold to the grocer. The primary concern of these coffee wholesalers was advertising. Since the tin containers are classified under the heading "advertising antiques," such bins are as collectible as those of more complex design and construction. The near-perfect condition of this Austin, Nichols and Company bin makes it a valuable container even though it is not as rare as the Lake Shore bin in photograph 32.

COFFEE AND TEA CONTAINERS

Tea and coffee were two of the many items available in country stores, and both were among the first goods to be packaged in metal containers. The East India Company shipped tea in metal-lined wooden crates, and pewter was often used to make oriental tea containers. The oldest container in this book (photograph 14) is a tea canister possibly made as early as the 1790s. It is an exceptionally large canister; tea containers were generally smaller, but more elaborate than coffee containers.

Coffee was packaged in tin containers as early as the 1840s and 1850s (photograph 17). Roasted coffee, however, did not become fashionable until about twenty years later. Before the 1860s, people bought green coffee beans and roasted them at home. Commercial roasting was also available, however, and manufacturers stressed the advantages of even roasting in their advertisements. The pioneers of packaged coffee similarly emphasized the evenness and economy of their pre-roasted beans. "Oh, I have Burnt my Coffee again," cries a housewife in an 1872 advertisement for Arbuckle's coffee. "Buy Arbuckle's Roasted, as I do, and you will have no trouble," advises her companion.

Early coffee containers were air-tight but not vacuum-sealed like today's coffee cans. This meant that the shelf-life of the coffee was not as long. Edwin Norton received a patent to pack under vacuum conditions in 1898, and coffee was available in vacuum-packed containers after the turn of the century.

41

41. The paper label on this brightly colored Rose Bud Coffee can is in near-perfect condition. The label is unusual because it features soft red roses and a bright white background. Few tin containers were originally white and fewer still remain white today. When tins are found in such condition, where and how they were stored through the years is often a mystery even to collectors. The tin was made by Ginna and Company in the 1880s or 1890s. The tin measures 7 inches (17.8 cm) by 6 inches (15.1 cm).

42, 43. Some collectors concentrate on a particular motif, and these tins would be an excellent addition to an Indian collection. Both Strong-Heart and Wak-Em Up coffee were packaged in containers made by the American Can Company. They could not have been manufactured before 1901, but the style of the two cans indicates that they are both quite early. The Strong-Heart tin is 6 inches (15.1 cm) tall, and the Wak-Em Up tin is 5 inches (12.7 cm) tall.

44. Collectors are delighted by both the copy and the designs of C.F. Blanke Tea and Coffee Company tins. The firm's owners called themselves "promoters of good goods" and featured a horsewoman holding a cup of coffee on the front of the tin. Blanke's was aware of the selling potential of tins with a secondary use. The coffee was sold in "fancy 2 lb. cans to give the consumer a set of handsome cans to retain for household use." This tin was made before 1900. It is quite rare; there are about one dozen known. It is 7.25 inches (18.4 cm) tall.

45. This three-pound Turkey coffee container is 10.5 inches (26.7 cm) tall and is stamped CANCO, the mark used by American Can beginning in 1912.

46. This beautiful tea container is a good example of the stunning differences possible with one-color lithographic printing. The tin was made by Ginna and Company in the 1880s. The background color is green, and the design and lettering are printed in black.

47. Advertising copy from the early twentieth century seems somewhat quaint today. "This coffee," reads part of the description on this Bridal Brand coffee container, "is grown where the soil, climate and irrigation are in the highest states of perfection and is imported and controlled by us." The tin was made between 1910 and 1920, and is 6 inches (15.2 cm) tall. Collectors are attracted to such a tin not only because of the colorful label, but also because of the unusual product name.

47

COCOA AND SPICE CONTAINERS

Spices, which today are readily available and taken for granted, were important, prized commodities in earlier times. Like coffee and tea, they traveled long distances before reaching the consumer and were easily damaged by the atmosphere. Metal packaging, therefore, was highly practical.

Most collectible spice containers were manufactured to hold several pounds of seasonings. Like the larger store bins, they functioned as shelf dispensers in groceries. Smaller home containers were a later development.

On the other hand, cocoa tins were produced for individual consumers. The powdered beverage was well-suited for packaging in airtight containers, but was not marketed in such a form until the beginning of the nineteenth century. A process was developed which removed most of the fat from partially processed cocoa beans (a cocoa bean is nearly fifty per cent fat.) The powdered product was a great deal more digestible and palatable than the earlier variety. (Cocoa has the distinction of being one of the earliest products associated with a trademark.)

Spice containers are not usually found as a separate collection but are part of a larger tin container collection. There are enough varieties of cocoa containers, however, that they are sometimes collected independently.

48. This Griffing's Lunch cocoa tin is extremely rare. It was made by Somers Brothers in Brooklyn, New York, possibly as early as the 1870s. It is a large cocoa tin—10 inches (25.4 cm) tall. The flowered border and central logo of Crave and Martin Company make it particularly attractive. Such complicated designs attest the skill of early tin lithographs.

49

49. According to Cleveland Chocolate and Cocoa Company, Rose's cocoa "induces sleep" and "aids digestion." A modern cocoa can make no such healthful claims. Pre-twentieth-century manufacturers were not restricted by federal labeling legislation however. Also, cocoa manufacturers tried to advertise the new, more palatable methods of cocoa processing as well as contrast their product to less-healthy coffee. The tin was made by S.A. Ilsley. The one-half-pound container measures 4.5 inches (11.4 cm) by 2.5 inches (6.4 cm).

50, 51. Although they are not as large as some store bins, these two spice tins were designed for grocery use. Each contained ten pounds of spice and would have been used as a dispenser. The lettering on both tins is printed in two colors. The pepper tin was made by Ginna and Company in the 1870s, and the allspice tin was probably made a short time later. The pepper tin is unusual for a Ginna tin because the design is transfer-printed. Well-known for lithographed tins, Ginna rarely used transfer-printing. The allspice tin has a paper label. Both cans measure 11 inches (27.9 cm) by 7 inches (17.8 cm).

52. Three small mustard containers show the English tendency toward attractive package design at the expense of the product name. All the labels are on the bottoms of the tins. The rectangular tin contained Moss, Rimmington and Company mustard, and the other two were filled with Farrow and Company mustard. The standing tins are each 3 inches (7.6 cm) tall, and the round tin is 2 inches (5.1 cm) tall.

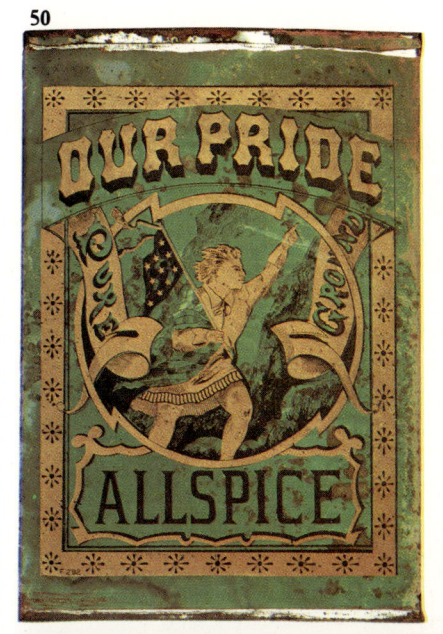

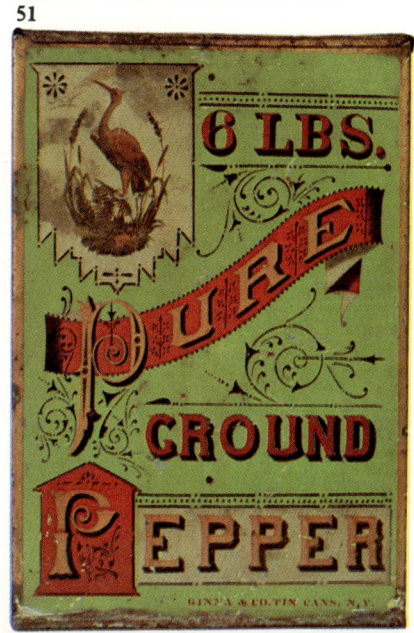

The Tin Can Book

53. Cocoa was a more popular drink in the first quarter of the twentieth century than it is today, and cocoa advertisements were geared to an older audience. The lady on this Lowney's cocoa container seems to be toasting her viewers.

54. "Physicians recommend cocoa to be nutritious and beneficial to the health," reads the label on Index cocoa. Montgomery Ward and Company further assures the public that an "analytical chemist" has fully tested its product. This tin was marketed after 1900, but before 1907. The passage of the first Food and Drug Act (1906) severely limited the claims a food producer could make on its labels.

55, 56. *La Belle Chocolatière* was taken from a painting by the French artist Ratoire, and it is the first trademark known to be associated with a product. The cocoa was available in tin containers during the last three decades of the nineteenth century, but the trademark was in use at least twenty years earlier.

57. The brand name Busy Biddy and the chicken on the front of this tin have little to do with spices. The tin was made in the 1920s. Its construction is a clue to its age: a small raised ridge keeps the lid from moving down the can. The tin is 3.5 inches (8.9 cm) tall.

58. This intricately decorated mustard tin was made by Ginna and Company for the firm of Griffiths, Griffin and Hoxie's Acme Mills Tea, Coffee and Spices of Utica, New York. The unusual Egyptian design demonstrates once again Ginna's facility with chromolithographic techniques. The tin was made in the 1890s, and measures 11 inches (27.9 cm) by 7 inches (17.8 cm). Like the pepper and allspice tins on page 53, it would have been used as a shelf dispenser in a store.

PEANUT TINS AND PEANUT BUTTER PAILS

Small peanut butter pails which originally contained from ten ounces to one pound of peanut butter are highly prized collectible tins. These colorful, rare miniatures were manufactured during the 1920s and 1930s to appeal to children. A half century later these little pails are still desirable because of this pleasing child-oriented design.

Peanut butter pails were often lithographed with brightly colored cartoon characters. The decorations and the size and shape of the containers made them ideal for toys. After the peanut butter had been eaten, they were given to the children. They disappeared in the sand-box, were misplaced or left behind. Consequently, few remain for today's collector. Pails with the original lid intact are the most highly valued.

The large five and ten-pound containers for salted peanuts are equally sought after. They were also manufactured during the 1920s and 1930s. Since they seem too large for home use, they were probably sold wholesale or used as dispensers in stores. They, too, are colorful and difficult to find. "Never pass up a peanut tin," advises Clark Secrest in *Tin Type*.

59, 60. Two colorful ten-pound peanut tins: both were manufactured during the 1920s or 1930s and are 10 inches (25.4 cm) tall. The top of the Squirrel tin suggests that the peanuts be placed in a glass dispenser for sale. Since glass shelf-dispensers were as much a part of the country-store era as peanut tins, the dating of the can is reconfirmed.

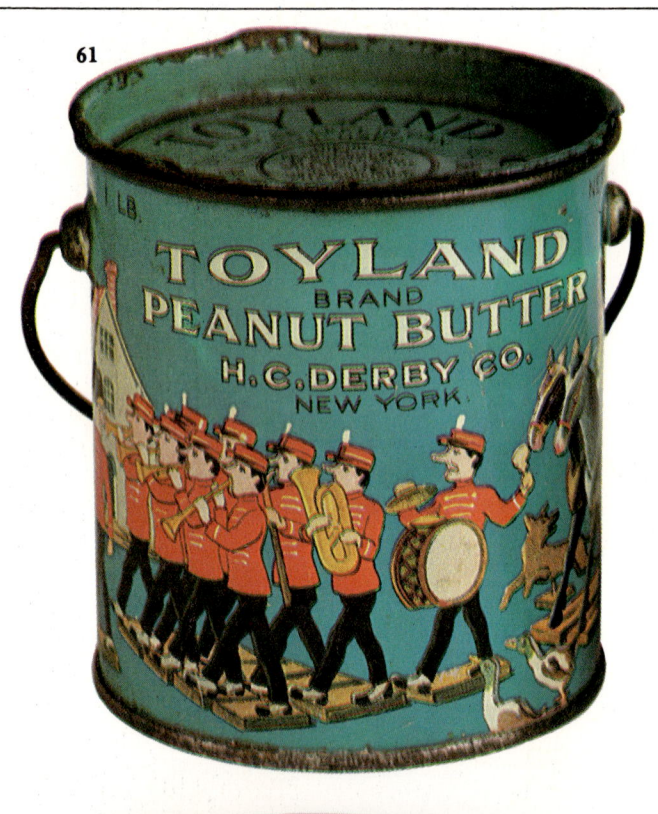

61, 62. These two little peanut butter pails—each about 4 inches (10.2 cm) high—are typical examples of this kind of container. The bright blue background of the Toyland pail and the many gay, primary colors of the Uncle Wiggily pail are unusual. The cartoons appeal equally to adults and children.

63. Jackie Coogan was a popular child actor in the 1930s. His endorsement of peanut butter undoubtedly helped to sell many little pailsful. The peanut butter was made by the Kelly Company of Cleveland, Ohio, and marketed under two brand names—Jackie Coogan and Dixie—both of which pictured Coogan on the pail. The reverse sides of the tins are decorated with cartoons showing different costumes and situations from Coogan's movies. Sets such as this are little collections within collections. They appeal to collectors, who attempt to find all known examples.

63

THE TIN CAN BOOK

64, 65. The Kelly Company not only made Dixie and Jackie Coogan brands peanut butter, but produced salted peanuts. These black and white Mammoth tins are favorites among collectors. The clear graphics and white background make the tins striking. Such detailed designs not only appeal to tin collectors, they also offer a clue to the age of the tins. These tins resemble those made by Ginna and Company and Somers Brothers before the turn of the century more than they do the Squirrel and Robinson Crusoe designs in photographs 59 and 60. The Mammoth tins were made by American Can Company, possibly before 1910, but certainly before 1920. The tin on the left held five pounds of peanuts and is 7 inches (17.8 cm) tall. The tin on the right held ten pounds and is 11 inches (27.9 cm) tall.

66. Superior Peanut company of Cleveland, Ohio, also handled both peanuts and peanut butter, under its Giant brand name. The large tins and the little pails were both decorated with a menacing, bearded figure complete with club. Like similar tins, this pail is 4 inches (10.2 cm) tall.

ENGLISH BISCUIT TINS

English biscuit tins are among the oldest collectible tins available to the collector. Huntley and Palmers were one of the first food producers to package their product in tin boxes, and as early as the 1850s they were featuring embossed and colored containers. Of the several biscuit manufacturers—Carr, Peek Frean, Jacob and Company, McVitie and Price, and the two dozen or so lesser known brands—Huntley and Palmers successfully dominated the market. As a result, most of the collectible tins were manufactured by the firm of Huntley Boorne and Stevens. (Huntley Boorne and Stevens did occasionally make tins for other biscuit manufacturers.)

For everyday use, the various firms made simple tins without elaborate decorations. But they outdid themselves for special occasions, especially Christmas. The tins must have been worth more than their contents even when they were made. Today some have been valued at several hundred dollars.

English biscuit tins cannot be found in as prime condition as American tin containers because of their greater age and the distances which they usually traveled. (Huntley and Palmers prided themselves on being known throughout the world.) But this does not lessen their value. These ingenious, imaginative containers probably bring more delight to collectors today than to those who found them beneath a Christmas tree.

THE TIN CAN BOOK

67

67. The manufacturers of English biscuits were fond of tins made in the shape of vases and urns. This particular Huntley and Palmers vase is unusual. Instead of the more typical dark brown background and somber illustration, the tin is decorated in the blue and white designs of Chinese porcelain. It was made in the 1930s, and is 6.75 inches (17 cm) tall.

68. This Huntley and Palmers tin from 1893 is decorated in a favorite style: it tells a story of English life in pictures. On the top a village blacksmith works with a horse. On the left side he is in church with his wife and children; on the right side the blacksmith and his eldest son are together at the anvil and forge; and on the back, the other three children hurry off to school. To amortize the cost of making dies for complicated shapes like this tin and the one in photograph 69, different pictures were printed on tins of the same construction.

69. The setting of this Huntley and Palmers tin is exotic Arabia. It is less a story tin than the Blacksmith tin on the left, but it still has a theme. On top a woman sips Turkish coffee. On the sides are scenes from the marketplace and of dashing Arabian princes on horseback. It was probably made before 1900. The designs are elaborately printed in many brilliant colors.

70. There are only two known examples of this Carr and Company commemorative tin celebrating Victoria's Glorious Reign from 1837 to 1897. Its theme is progress. The top shows traveling in 1837 and in 1897. On the front is the Forth Bridge—the longest ever constructed at that time—and the Cunard steamships—the first to cross the Atlantic. On the back are shown the telephone, the telegram, and the conquest of India.

THE TIN CAN
BOOK

68

69

70

63

THE TIN CAN BOOK

71. Competition among the makers of Christmas tins for English biscuits was fierce. Each was more elaborate and complicated than the next. It is impossible not to be charmed by this working Coronation Coach which once held W. and R. Jacob and Company biscuits. The tin was made in 1936. Because the British manufacturers produced superior tins until the 1940s, dates are somewhat less crucial than they are to other types of collections. A superior 1940s tin (one that is in good condition and attractively made) is equally as collectible as one made before the turn of the century. The coach is 5 inches (12.7 cm) tall and 9 inches (22.9 cm) long.

72. This Grandfather's Clock, made for Huntley and Palmers in 1929, is one of the most beautifully decorated English biscuit tins. The tin is black, and the images are printed in silvery colors. It is 11.5 inches (29.2 cm) tall.

73. Of course, the blades of Huntley and Palmers' Windmill biscuit tin turn. When most people see these tins for the first time, they ask, "Where were the biscuits?" The rounded top behind the blades opens to dispense the biscuits. The woman in the doorway is holding a standard package of Huntley and Palmers biscuits. This tin was made in the 1920s and is 16 inches (40.6 cm) tall.

64

THE TIN CAN
BOOK

74. This spinning top is another masterful Huntley Boorne and Stevens creation for Huntley and Palmers. The number of pieces and the shaping involved in making such a tin are fully as remarkable as the complicated embossed designs on the facing page. The top, of course, actually functions. It is 8.5 inches (21.6 cm) tall.

THE TIN CAN BOOK

75. This set of Huntley and Palmers' books was made in 1901. Books were a popular design—both single volumes and sets like this were filled with the biscuits of various English bakers. The little tab on the top of the books lifted the tin's lid. Complicated shapes like this involved a good deal of hand soldering. These are small volumes; the tin is 6.25 inches (15.9 cm) tall and 6 inches (15.2 cm) long.

76. The firm of Huntley Boorne and Stevens did beautiful tinplate work for Huntley and Palmers. These three tin purses are amazing in both texture and color. The purse in back is embossed with an elephant caravan on the facing side; on the reverse side, a tiger hunt is in progress. The Indian image is complete with a small handle in the shape of an elephant. The purse on the left, made in 1902, is a colorful rendition of a scene from the animal world. A bright green lizard hunts a dragonfly. The lizard and insects are embossed, as are the ridges and whorls in the log on which they perch. The purse on the right appears woven, and the texture of the pastel-colored metal duplicates the effect perfectly. It was made in 1901. This ability to texturize metal is one of the fascinating skills of the makers of English biscuit tins. A one-color embossed tin can be as exciting as a printed one. The careful coloring adds another dimension.

77. Another outstanding Huntley Boorne and Stevens tin made for Huntley and Palmers: this old-fashioned telephone is complete in every detail. At first glance, it does not appear to be a biscuit tin at all. The little telephone is even more intriguing because it is only 8 inches (20.3 cm) tall.

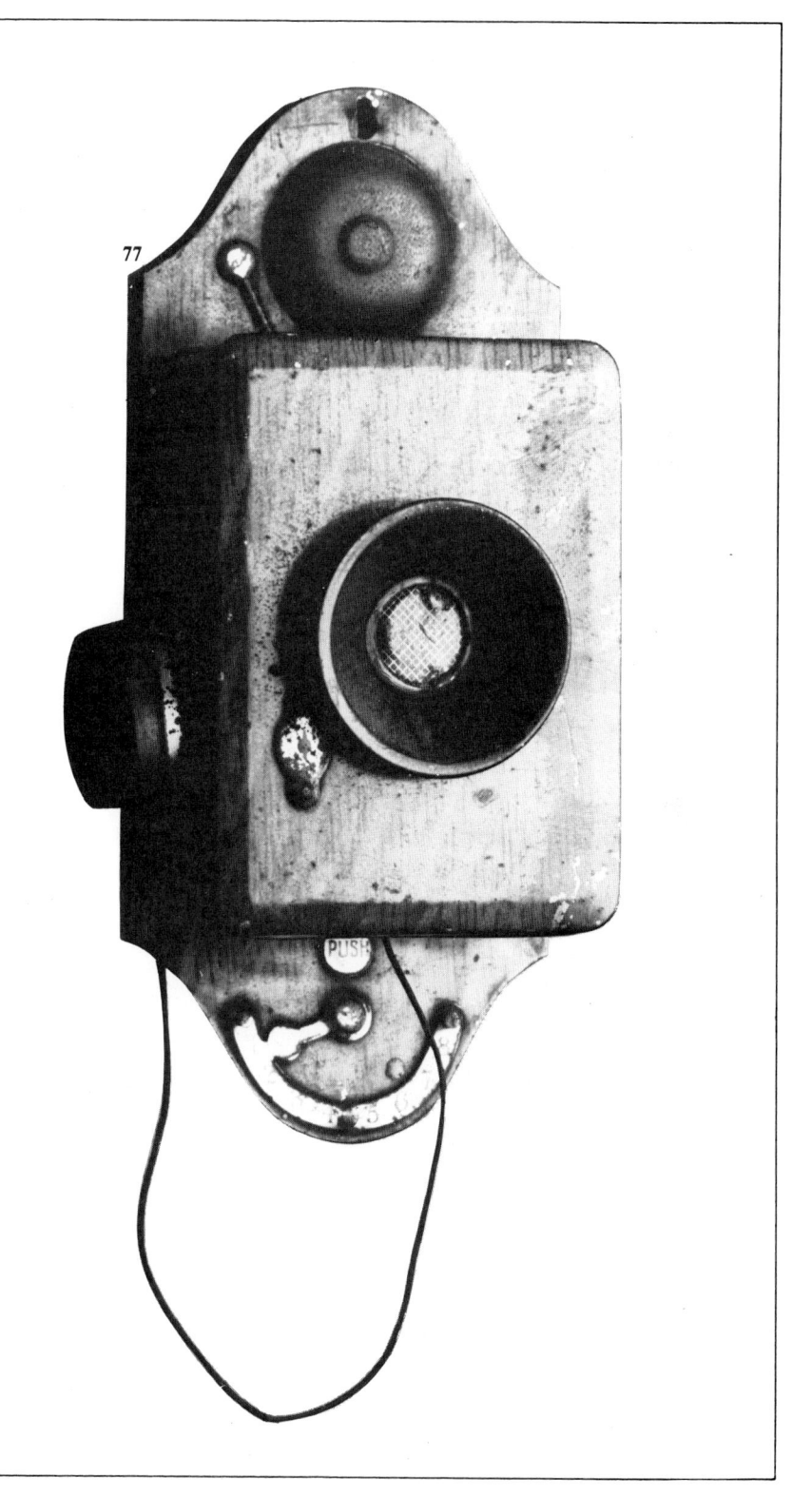

VEGETABLE TINS

Hermetically sealed vegetable tins account for the greatest number of tin cans produced in the world. It was vegetable and petroleum cans that first gave the tinplate and can manufacturing industries the impetus toward mechanization which ultimately resulted in the production of the decorated metal box. A general tin container collection is probably not complete without an example of the hermetically sealed can, but vegetable tins seldom form the basis of an entire collection. (Beer cans, of course, are the twentieth century equivalent of early hole-and-cap vegetable cans.)

The lowly vegetable tin is neglected purposely. Unlike the decorated tin container, it was designed to be discarded. No matter what method was used to open the can, its life expectancy was short. Few remain in reasonable condition. The occasional discovery of a bruised and rusted can with a clear and clean label is cause for suspicion: the label is very possibly a later addition.

However, vegetable tins in excellent condition can be found, and the detail and artistry of their lithographed paper labels have the same appeal as the printing on any other tin container. (Lithographers once proposed printing labels directly on vegetable tins, but the canning industry never adopted their suggestion.)

THE TIN CAN BOOK

78, 79. Labels were more idyllic in the nineteenth century. These two are designed to elicit a positive response from the customer rather than to broadcast the brand name, or even the product inside. On the left on a label for Sweet Green Beans a cherubic little boy plays with empty tin cans. On the right a cook wearing striped socks tastes his own Epicurean mock turtle soup. The cans are the same size as a modern food tin.

80. H.J. Heinz as well as the Libby brothers and their partner McNeil; Joseph Campbell; and the Biardots, founders of Franco-American Food Company, were pioneers of the canning industry. This pail of Heinz plum butter from the early twentieth century is a particularly attractive tin. Paper labels similar to this one were also applied to crocks of vegetables.

69

THE TIN CAN BOOK

81
82
83
84

81-84. The labels on these four tomato tins tell a story: 79. A. Anderson was a New Jersey tinsmith who opened a cannery in the 1860s. 80. In 1869, he went into partnership with Joseph Campbell. Campbell wanted the business to enlarge; Anderson was content with a modest enterprise. 81. In 1876, the partnership was dissolved when Campbell bought Anderson out. 82. In 1880, Anderson started again and opened his own cannery. But Campbell went on to bigger and better things – soup.

85, 86. Gail Borden is one of the heroes of the tin-can era. His condensed milk, first marketed in 1856, almost single-handedly established canned goods as part of American diet. This particular can was probably manufactured during the early 1900s. The only date mentioned on the label is 1899, and the can is constructed like the hole-and-cap cans of that era. (By the 1910s the sanitary can was beginning to overtake the market.) Condensed-milk cans were smaller than ordinary cans, as they are today. The original can is shown on the trade card in photograph 10. This one is 3.5 inches (8.9 cm) tall.

THE TIN CAN BOOK

74

89. Educator Cakelets were made by Johnson Educator Food Company of Cambridge, Massachusetts. Dr. William Johnson, the creator of Educator crackers, was a Boston dentist. He was convinced that white bread and other refined foods were ruining his patients' health and decided to experiment with a nutritious, whole wheat product in a convenient form. He baked his first crackers in 1886, and was soon out of the dental business. In 1891, Dr. Johnson bought Butler's Bakery in Newburyport, Massachusetts. In 1910, contrary to Dr. Johnson's original intention, the firm began to sell sweet crackers. Because this Educator tin contained cakelets, it must have been made after 1910. The tin was made by Tindeco and measures 3 inches (7.6 cm) by 6 inches (15.2 cm).

90-92. The cocoanut tin in the center was made by S.A. Ilsley; the other two are Ginna tins. Therefore, all three were made during the 1880s or 1890s. On the back the Ilsley tin is decorated whimsically with an outdoor scene involving a hunter, a lutist, and a rabbit. The others, depicting monkeys and cocoanuts, are more traditional. They are 5 inches (12.7 cm), 5 inches (12.7 cm) and 6.25 inches (15.9 cm) tall, respectively.

93, 94. This spectacular Home Made Ginger Wafer tin dates probably from the very early 1900s. (Clues to the age of the container are the clothing and hair style of the girl pictured on it.) It is 9.25 inches (23.5cm) tall. The container is made from tinplate but the label is paper. Such labels were applied after the tins were filled. The paper not only identified the container, but also effectively sealed it. It was necessary to rip the paper to open the can.

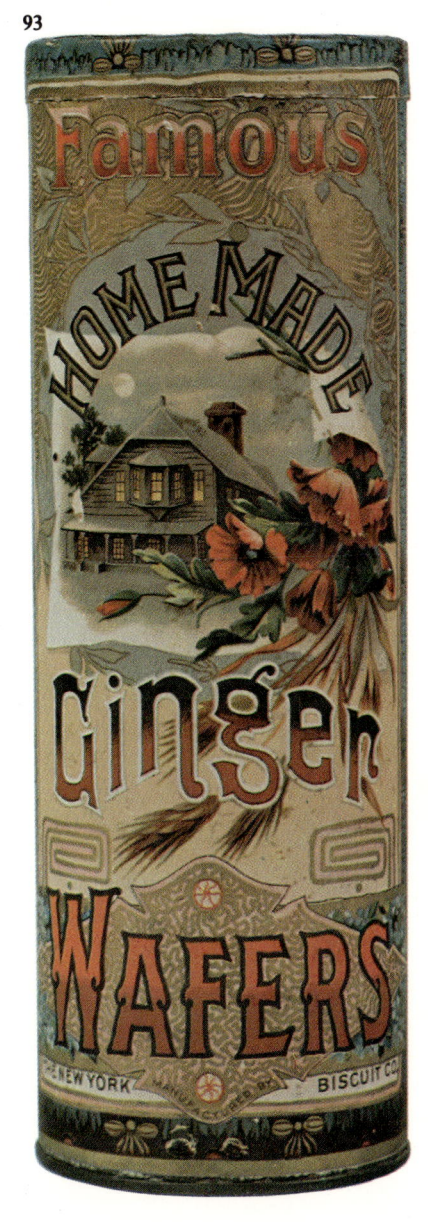

THE TIN CAN BOOK

95. Collectors consider this Arm and Hammer lunch pail a particularly strange container: a baking-soda lunch pail is an oddity. The tin held ten pounds of baking soda and is 7 inches (17.8 cm) high and 8 inches (20.3 cm) long. The logo is dark blue, and the background color a dark yellow. On the reverse of the tin, the logo is red and reads "Arm and Hammer Baking Soda." The tin was made by Tindeco in the 1920s or 1930s.

96, 97. Metal packaging was so pervasive that even chewing gum came in tins. These two date from the early 1900s.

98, 99. P. J. Towle, a grocer from St. Paul, Minnesota, first produced Log Cabin syrup in 1887. Distressed by the high cost of maple syrup, he mixed maple and cane sugar syrup to make a cheaper but good-tasting blend. The syrup—and the little log cabins in which it was packaged—became so poular that The Log Cabin Products Company became part of General Foods in 1927. The one pound tin on the left, 3.5 inches (9 cm) by 4 inches (10 cm), is the older of the two shown. It has a copyright date of 1914, and was made in St. Paul. The family size tin on the right, 5 inches (12 cm) by 5 inches (12 cm), was a product of General Foods.

100. The Yellow Kid was the creation of New York cartoonist Richard Outcault in 1894. In the comic strip, the Kid's reactions to his urban situation were always written on his yellow gown. It was the first comic strip printed in color and eventually made Outcault famous. The container stands 13.5 inches (34.5 cm) high and was made by S.A. Ilsley. It was made, therefore, before the turn of the century. It is the only known Yellow Kid tin of its kind and one of the largest Ilsley tins extant. The label is paper.

76

The Tin Can Book

TALCUM POWDER TINS

Talcum powder tins have recently become popular enough to warrant their own category. Their appeal is not difficult to understand. They are decorated with smiling babies and lovely ladies, and colored with pastel hues not often seen on other types of tin containers.

Gerhard Mennen of Newark, New Jersey, first marketed talcum powder in a pasteboard drum in 1890. The cardboard leaked and Mennen commissioned Somers Brothers to design a tin container. The Somers tin had a cap fitted over a revolving sprinkler top which, when turned, permitted the container to be alternately opened or sealed. A later version had a small piece of metal inside so that after the powder was gone, the container could be used as a baby's rattle.

Mennen advocated using his borated talcum powder in the nursery, but other brands were recommended as relief for a large variety of minor illnesses. Talcum powders were also sold as perfumers and face powders. Sometimes medicinal and cosmetic properties were combined.

Most collectible talcum powder containers were manufactured during the first quarter of the twentieth century. Later designs became progressively less interesting and metal finally gave way to plastic.

In general, talcum powder tins are not as expensive as some other collectible tins because they have not been popular as long.

THE TIN CAN BOOK

101. The Somers Brothers made Gerhard Mennen's first tins in the 1890s. However, this one was made by American Can Company. Since there is no mention of the Food and Drug Act of 1906 on the tin, it was probably made between 1901 (when American Can was formed) and 1906. The design is the same as on the original Somers can. Pictured on the front is the child of one of Mennen's employees. On top is the venerable Mennen himself. The baby was appealing, and Mennen's portrait was meant to inspire confidence. The tin is 4 inches (10.2 cm) tall.

101

79

THE TIN CAN BOOK

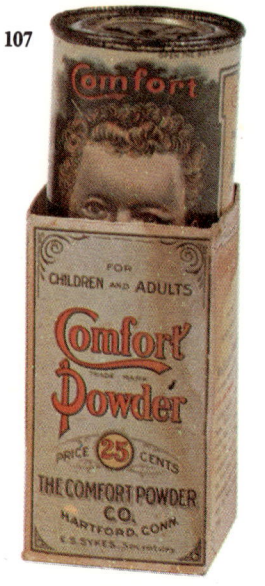

102-104. Colgate's baby talc was available in two sizes; the tiny tin on the right is a sample. Sets like this one, with samples and assorted sizes of the same product, are highly collectible. They offer a focus for the treasure hunt. These tins could have been made as early as 1910, and were definitely made before 1920. The tallest is 6 inches (15.2 cm). The others are 4.5 inches (11.4 cm) and 2 inches (5.1 cm) tall, respectively.

105-107. The manufacturers of Comfort powder attempted to make their market as broad as possible by advertising a variety of potential uses. The Comfort powder in the tin on the left is "unequaled face Powder." The smaller tin on the right combines all three possibilities in the phrase: "The Modern Nursery, Toilet and Sick Room Medicated Powder." The two larger tins, 4.25 inches and 4 inches (10.8 and 10.2 cm) tall, respectively, sold for fifty cents. The smaller tin, 3.75 inches (9.5 cm) tall, sold for twenty-five cents. The package must have been more expensive than the product; the box is as elaborately lithographed as the container. Comfort tins have recently become so popular that some people collect them exclusively. There are as many as twelve to fifteen different container styles.

108. This Vantine's talcum powder tin is the most ornate of all talcum powder containers. It is relatively hard to find and is sought after as a collectible. Because the Vantine's tin is desirable, it offers collectors a challenge. Such a tin may not be the favorite, however. Collectors tend to seek things they like, and the baby in the picture is less appealing than the Colgate child or the Mennen infant. The tin was made in 1925, and is 5 inches (12.7 cm) tall.

109. The Allen Pharmacal Company made Rose de France borated talcum powder as well as Violet borated powder. The various talcum powder producers hoped, of course, to sell more powder by manufacturing different scents. "Borated" simply meant mixed or impregnated with boric acid, a mild antiseptic.

110. Like the manufacturers of Comfort powder, the makers of Perfumed Violet talcum powder also stressed the many uses of their product. It was "For the Toilet of Babies and Adults." It not only "Softens and preserves the Skin," but "May be used also as a Face and Tooth Powder with pleasing result."

111. The man on the front of Fehr's compound powder is "The late Dr. Julius Fehr." Mennen had his own picture on top of his talcum powder tins. It was thought that such distinguished-looking gentlemen helped to generate consumer confidence. Fehr's powder company was located in Hoboken, New Jersey. The tin was made during the first quarter of the twentieth century.

112. Lovely ladies graced talcum powder tins more often than babies. Containers decorated like this one were obviously designed to be sold to women as toilet powder. The manufacturer concentrated on a segment of the market rather than stressing a variety of uses for his product. This Corsage Bouquet tin was made before 1920. Like the other containers pictured on this page, it is 4.5 inches (11.4 cm) tall.

113, 116. Winchester and Reel Man are two of the few early talcum powders marketed for men. Both tins are from the 1920s, and are 4.75 inches (12.1 cm) tall.

114. Using magazine advertising of the day as a guide for design, this Jess talcum powder was available from 1909 until 1919. They are among the loveliest talcum powder tins. The softly colored designs are embossed, which adds additional beauty. The embossing also adds value to the tin; few American containers were embossed.

115. During the early decades of the twentieth century, perfumed talcum powder in containers designed especially for women were in great demand. Since the powders were all fairly similar, package design was one way for the manufacturer to make the product seem unique. In addition, advertising on the Cloverine package claims that it is different from other powders, but does not say how.

83

COSMETIC AND PHARMACEUTICAL CONTAINERS

Cosmetic containers, particularly the powder tins, are some of the prettiest collectible tins. There are few cosmetic specialists, however, and like pharmaceuticals, these tins are generally part of a larger collection.

Airtight tin containers were a boon to the pharmacy trade. Pharmacists began grinding chemicals in large lots by the 1850s, and production generated a need for mass distribution. Tin containers were the perfect package. At first seamless tins were used, but as manufacturing techniques improved, more complicated designs developed.

Elaborate construction, however, was not the forte of the makers of pharmaceutical containers. For the most part, can manufacturers produced stock shapes, printed with the graphics desired by the pharmaceutical firm. Consequently, pharmaceutical tins show some of the more interesting graphics on tin containers—in content if not in form. A fascinating array of products—cures for cancer, rheumatism and arthritis; salves composed of unknown ingredients and of uncertain benefit; powders curing everything from sore feet to skin rashes—are packaged in tin. Elaborate claims recall the patent-medicine era of which they were a part. (In fact, the earliest known labels were made by seventeenth century European vendors of quack medicines.) The passage of the first Food and Drug Act in 1906 restrained to some extent the exuberant sellers of non-prescription drugs.

117. Although this face-powder container was made in France, the Cafe Martin was a New York establishment. It may have been a souvenir item, given away by the proprietors of the cafe to their patrons. American collectors obviously do not have much access to foreign containers and they are highly prized. A tin as attractive as this one would be collectible no matter who made it. The container has a hinged lid that snaps shut. The tin is 3 inches (7.6 cm) in diameter.

118. This early Smith Brothers tin was made by Ginna and Company. Therefore, it must have been manufactured during the 1880s. Though the tin is quite small, the lithographed portraits of the brothers are quite detailed. It measures 3.75 inches (9.5 cm) by 2.25 inches (5.7 cm) and sold for ten cents.

THE TIN CAN BOOK

119, 120. Mrs. Dinsmore's and Dr. White's cough-drop tins were both made by Somers Brothers as early as the 1880s. Collectors value the tins because of the quality of the graphics. Both tins are excellent examples of the possible detail using lithographic printing. Collectors also appreciate the old-fashioned advertising copy. Dr. White's Cupid brand cough drops, say the label, "have a most agreeable taste and can be eaten freely without the slightest unpleasant effect, they are in truth a veritable delicacy." Not only that, they "have become most popular with public speakers, singers, actors, and professional people in general who find them of great value for clearing and strengthening their voices." The tins are approximately 7 inches (17.8 cm) tall.

121, 122. Moses' Celebrated cough-drop tins were made by Somers Brothers in the 1880s. The cough drops were manufactured by E. J. Hoadly of Hartford, Connecticut, and "EJH" was stamped on each drop. Moses tins are quite rare. Both shown are orange. The full-sized tin is 7 inches (17.8 cm) tall, and the sample is 2.25 inches (5.7 cm) tall.

123. This tin of hair grower was probably manufactured before 1906. Although the product may have controlled dandruff, it could not have promoted hair growth. Such claims could not have been made after the passage of the Food and Drug Act of 1906. The tin is 3 inches (7.62 cm) square and 1.5 inches (3.8 cm) deep.

124. J.J. Carnaud made the container; M.J. Botot made the cake of tooth powder packaged inside. European containers such as this are rarely available to the American collector and are highly prized. The advertising in French makes this tin more obviously foreign than the Cafe Martin tin in photograph 117. The tin is 2.75 inches (7.0 cm) on a side.

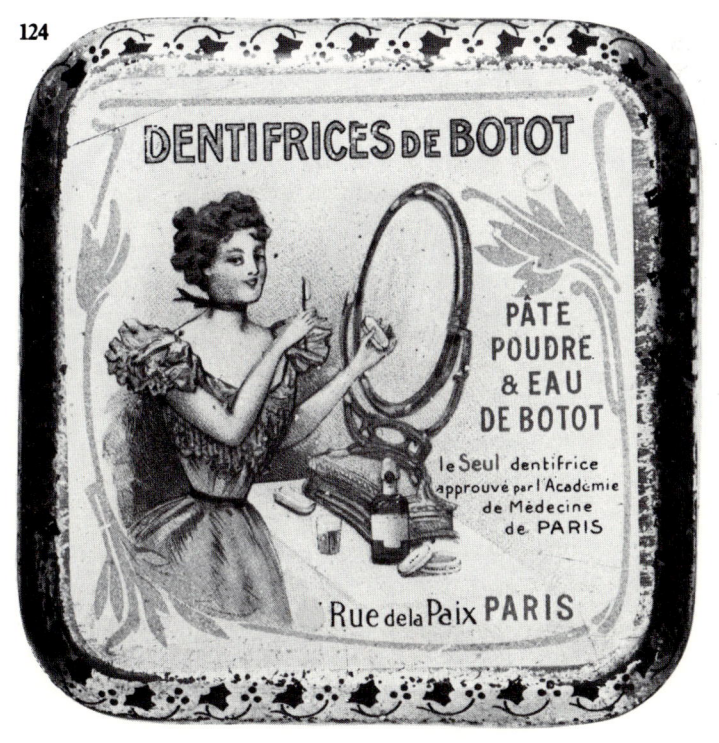

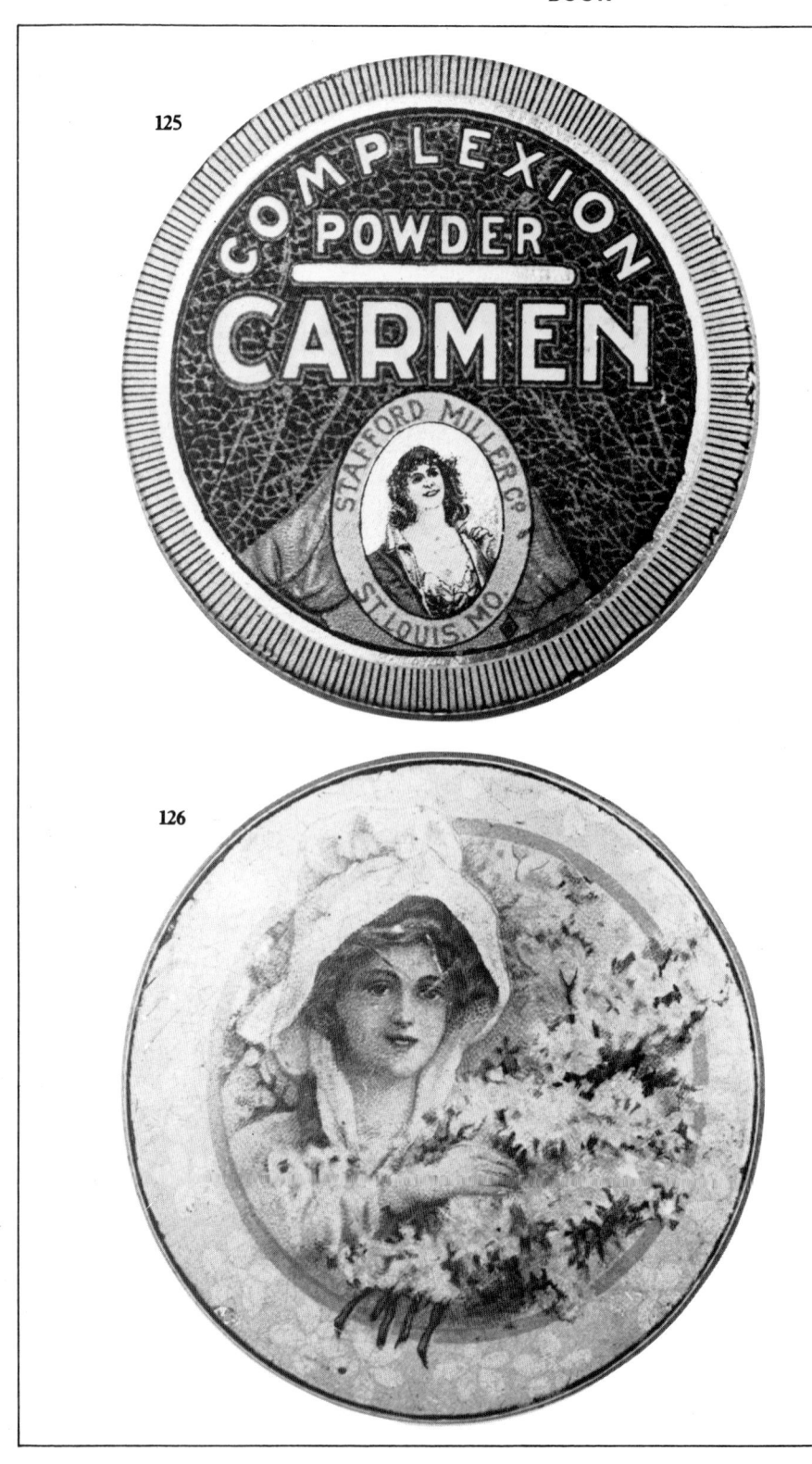

125, 126. Both of these face-powder tins have a small mirror on the back. The lower tin contained Fair Maid complexion powder. The inside of the tin is decorated with the same design as the top, but would have been visible only after the powder was gone. Both tins were available in the 1910s.

127, 128. Achille Brioschi and Company, Incorporated, was located in Milan, Italy, but its product was made in the United States. Two tablespoons in half a glass of water was sufficient for relief of indigestion. Like so many late nineteenth- and early twentieth-century corporations, the firm is still in existence today.

129. Peerless tooth powder was made by the American Stopper Company of Brooklyn, New York. The firm became part of American Can in 1905-1906. Certainly, it is more attractive than most modern toothpaste tubes, but the advertising is much less direct. The picture on the front of the tin would not immediately indicate that the product is tooth powder, even though the lady is smiling. It is small, only 3 inches (7.6 cm) tall.

130. Henalfa, "The Perfect Hair Restorer," was made from "Pulverized Henna Leaves and Herbs." It was "Guaranteed Harmless" and promised a "Uniform Shade." The tin is unusual. Cough drop and salve tins are easier to find than containers of hair preparations. The label gives an insight into the era. Though Henalfa is obviously hair dye, nowhere on the tin is the word "dye" or "color" mentioned.

THE TIN CAN BOOK

127

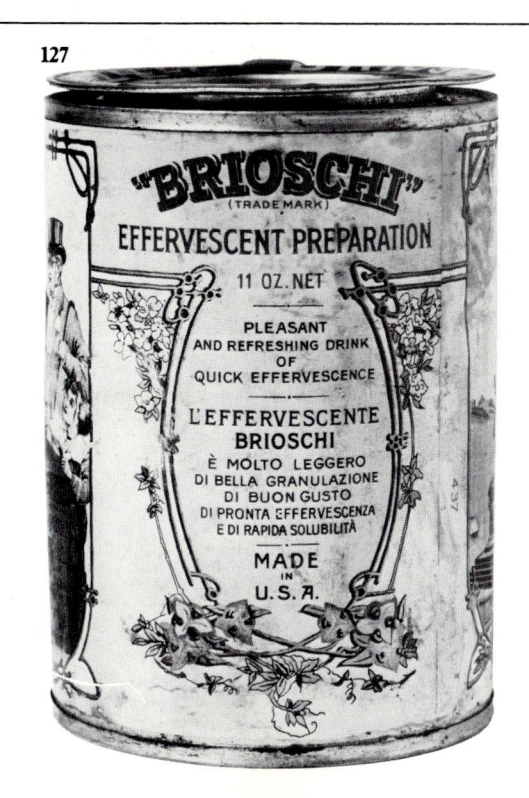

"BRIOSCHI" (TRADE MARK)
EFFERVESCENT PREPARATION
11 OZ. NET
PLEASANT AND REFRESHING DRINK OF QUICK EFFERVESCENCE

L'EFFERVESCENTE BRIOSCHI
È MOLTO LEGGERO
DI BELLA GRANULAZIONE
DI BUON GUSTO
DI PRONTA EFFERVESCENZA
E DI RAPIDA SOLUBILITÀ

MADE IN U.S.A.

128

129

PEERLESS TOOTH POWDER

130

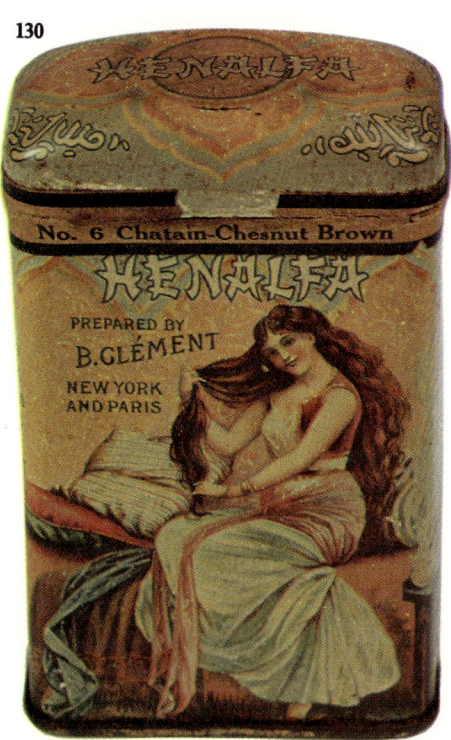

No. 6 Chatain-Chesnut Brown
HENALFA
PREPARED BY
B. CLÉMENT
NEW YORK AND PARIS

89

TOBACCO CONTAINERS

Tobacco tins are the most collected tin containers. They are also the most numerous. Literally thousands of different brands of tobacco were marketed during the early part of the twentieth century, and many of them were packed in boxes. Pocket tobacco tins were as commonplace then as cigarette packages are today. The reason was simple: tobacco is easily damaged by the atmosphere, and a tin is not only durable but airtight.

Several varieties of tobacco containers developed. One type was the flat pocket tin—a small, rectangular box with a hinged lid. Much more numerous are the standard pocket tins. Often, but not always, they are slightly concave or kidney-shaped to allow them to fit comfortably in the back pocket of a pair of trousers. Most have an embossed area on the bottom on which a match may be struck.

Tobacco lunchboxes are nearly as numerous as pocket tins, and are certainly as collectible. They were manufactured from 1901 until about 1925, and there are as many variations of the basic rectangular handled shape as there are brands of tobacco. The lunchbox design was a highly successful advertising gimmick—adults and children alike carried their lunch in tobacco boxes during the early twentieth century.

Tobacco canisters which held both cigars and cut tobacco are often beautifully lithographed in many colors. Though the quality of the canisters makes them highly collectible, they are the least numerous and therefore the least often found tobacco tins. Flat cigarette tins are another smaller group of tobacco collectibles.

131, 132. Richard Felton Outcault, the same cartoonist who created the Yellow Kid label in photograph 100, drew these paper labels for Buster Brown cigars. Pictured on the containers is Buster himself, his dog Tige, and a talented gentleman who is able to blow smoke out of his ears. These two are the only known examples of the canisters. The tins were made in 1926, and are 5 inches (12.7 cm) tall and 5 inches (12.7 cm) in diameter.

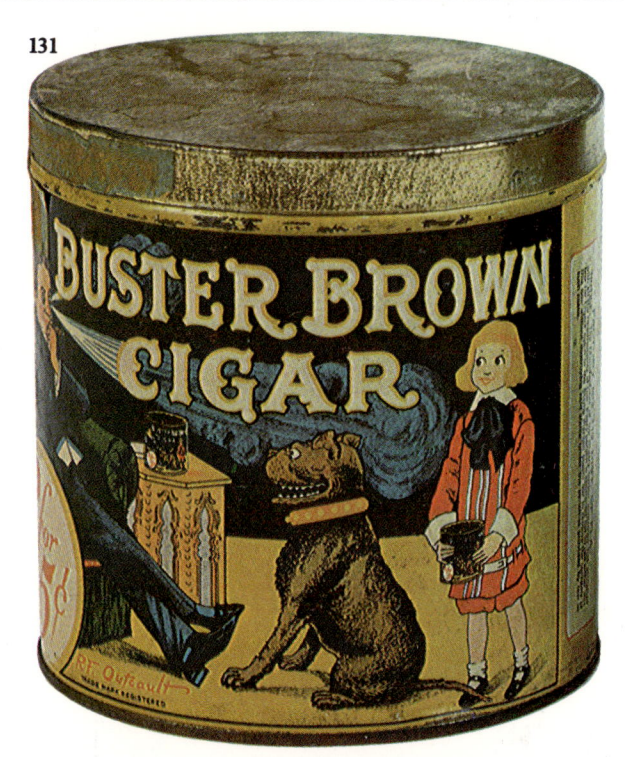

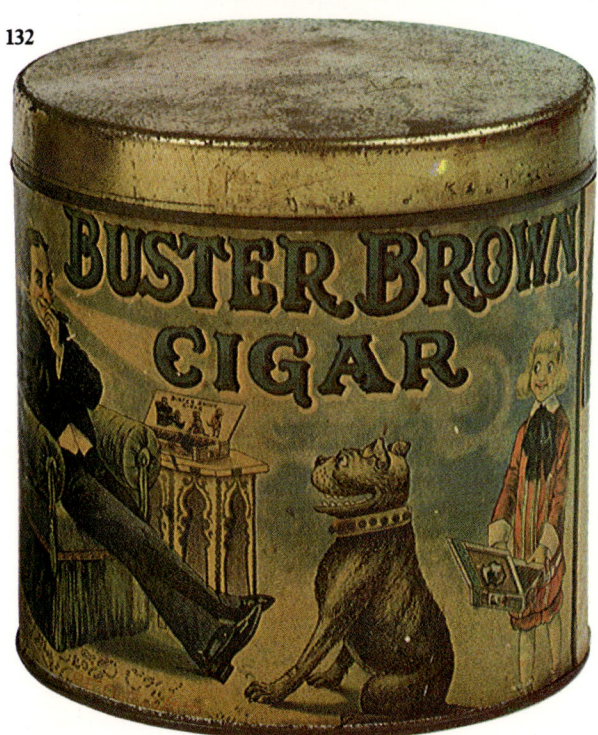

THE TIN CAN BOOK

133. His Master's Choice cigars were also packaged in a tin container with a paper label. This is an especially old cigar tin, made as early as 1870. It is 5 inches (13 cm) tall and 4 inches (10 cm) in diameter.

134. Scotten, Dillon Company of Detroit, Michigan, maker of Long Distance chewing or smoking tobacco, was one of the largest tobacco manufacturers in the United States. Daniel Scotten, rising from apprentice to tobacco mogul, founded the largest of the three big Detroit tobacco firms. (Globe and J.J. Bagley are the other two.) This tin has a CANCO stamp and is 6 inches (15.5 cm) in diameter.

135-138. Any collector would be ecstatic about owning all four of these Continental Cubes tobacco containers. Not only are the graphics some of the most beautiful printed on tin, but the sliding tops of the three standing tins are unique to Continental Cubes. The kidney-shaped canister in photograph 135, made like an enlarged pocket tin (7.5 inches, 19.1 cm) is extremely rare. Even more difficult to find is the canister in photograph 138 (5 inches, 12.7 cm), and the smaller kidney-shaped canister in photograph 136 (6 inches, 15.2 cm). The 4-inch (10.2 cm) pocket tin in photograph 137 is more common. A set like this one, of course, makes each individual tin more valuable. All of the tins were manufactured between 1905 and 1915.

The Tin Can Book

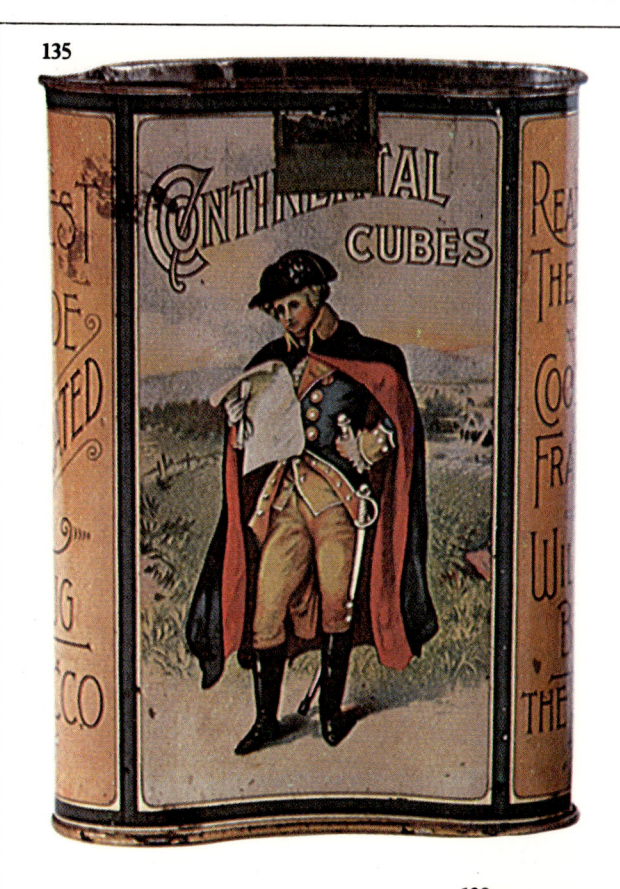

135

136

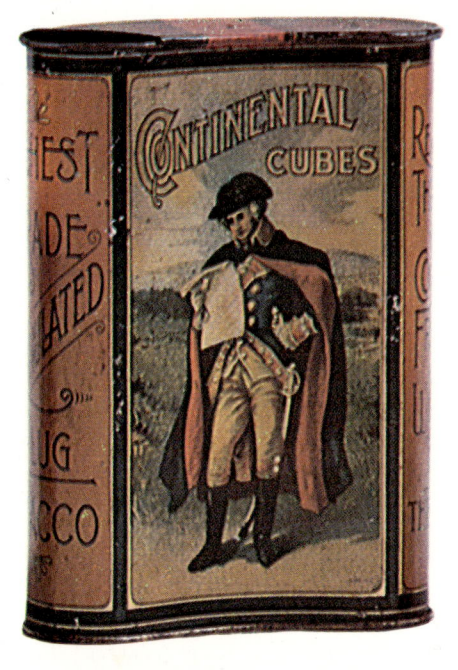

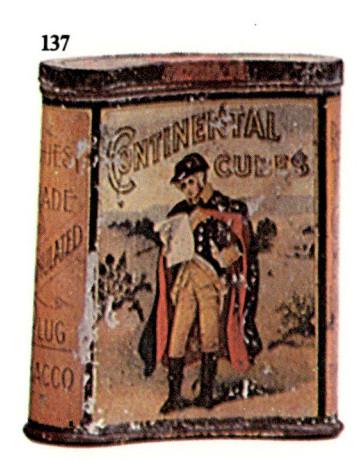

137

138

93

THE TIN CAN BOOK

139-144. All six of these small flat pocket tins, 3.75 inches (9.5 cm), by 2.25 inches (5.7 cm) are very old and hard to find. They were all made in the 1870s and 1880s. The Century tobacco container (photograph 140) was made by Somers Brothers. It features a revolving plate under a window (the plate is shown between pictures) with portraits of 1880 presidential candidates Chester A. Arthur and James Garfield, and their opponents Winfield Hancock and William English. The Old Glory, S.F. Hess and Co., and Florida fine-cut tobacco tins (photographs 141, 142, and 143) were all made by Ginna and Company. The May Queen tobacco container (photograph 144) was made by The Chicago Stamping Company; the North Star tin (photograph 139) was a product of S.A. Ilsley and Company. All are excellent examples of late nineteenth-century one-color lithography on tinplate.

145-148. These three Dixie Queen tobacco canisters (the package in photograph 145 is paper) were made over a period of twenty years. The larger canister in photograph 146 is a prize in the world of tin containers. It is marked Ginna and Company Tin Cans and has a paper label. The pail in photograph 147 has a 1902 tax stamp and is embossed above its paper label with the words "Smoke Dixie Queen Plug Cut Tobacco." The smaller canister (photograph 148) is the least colorful and most common of the three. This one has a tax stamp dated 1910. They are each approximately 6 inches (15.2 cm) tall.

THE TIN CAN BOOK

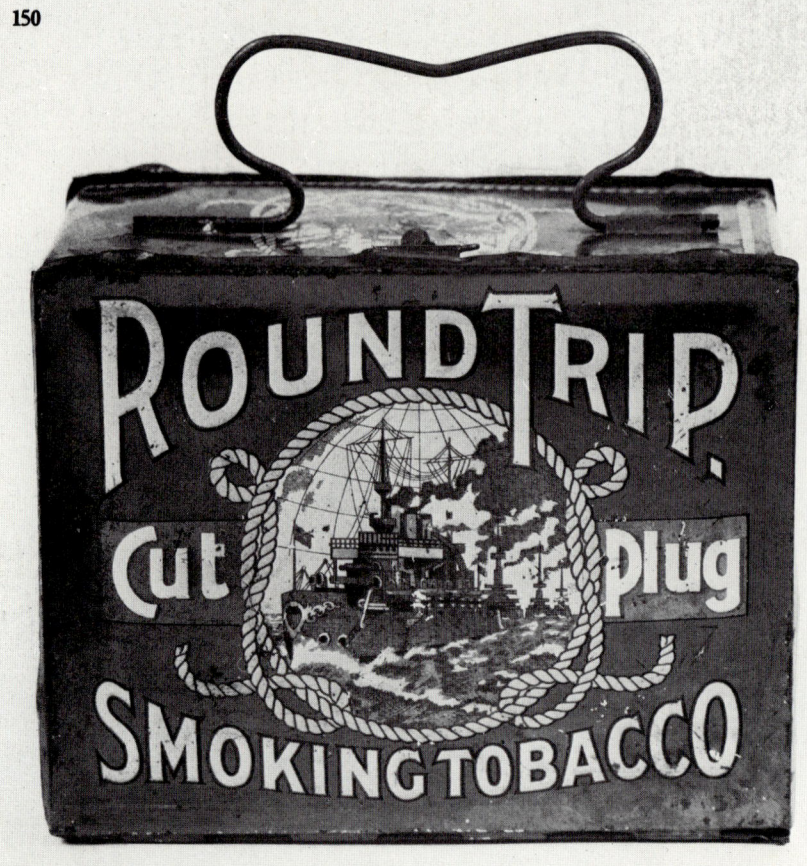

149. Some flat pocket tins, like this Prune Nugget container, had square corners. The fruit-flavored tobacco was made by Harry Weissinger Tobacco Company, Inc., of Louisville, Kentucky. The tobacco tax stamp on the bottom of the tin indicates that it was made as early as 1893. It measures 6.25 inches (15.9 cm) by 3 inches (7.6 cm) by 1.5 inches (3.8 cm).

150. This Round Trip cut-plug tobacco lunch pail has a 1910 tax stamp and was made by the American Can Company for Larus and Company tobacco manufacturers. Tax stamps were sometimes purchased by the tobacco manufacturers in advance, and a given issue could be in effect for as long as twenty years. They are not, therefore, an accurate means of dating a container, but they do indicate a general era date. A "plug" is any form of pressed or twisted processed tobacco used primarily for chewing. Such plugs were shredded or cut for use as smoking tobacco. Cut-plug tobacco could be smoked in a pipe, rolled in cigarette paper, or pinched and chewed. This lunch pail measures 6 inches (15.2 cm) across.

151. This Lime Kiln Club tobacco tin is the only one of its kind known. J. J. Bagley and Co. was another of the large Detroit tobacco firms. John Bagley, like Daniel Scotten, began as an apprentice in the Miller tobacco factory and went on to found his own prosperous organization in 1853. John Bagley died in 1881, but the company continued under his name until it closed in 1923. The paper label on this particular container is translated into four languages—English, French, Spanish, and German. It was made by American Can Company and is 5.5. inches (14.0 cm) tall and 5.25 inches (13.3 cm) in diameter.

151

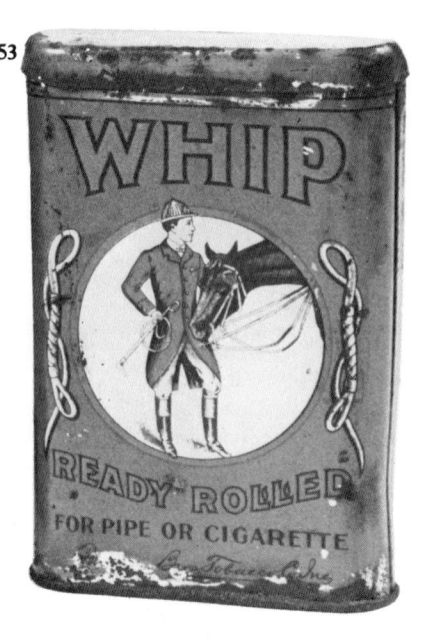

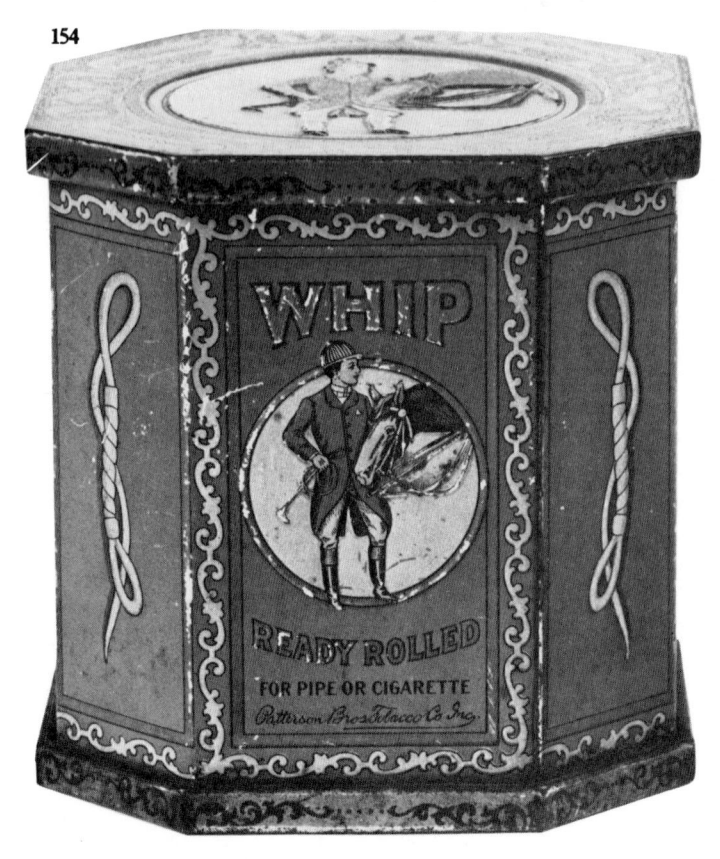

152. Heekin Can Company made this softly colored Brown Beauties cigar tin. Heekin was one of several independent manufacturers which opened after American Can Company was formed in 1901. The can is 5.5 inches (14.0 cm) tall and 3 inches (7.6 cm) wide.

153, 154. This Whip tobacco canister (photograph 154), shown with a matching pocket tin (photograph 153), is unique for two reasons. It is hexagonal, and the designs are embossed. (A different design appears on the back: a horseman holds a whip which cuns over his head to spell out "Whip.") Whip tobacco was produced by Paterson Brothers Tobacco Company. The canister is 5.5 inches (14.0 cm) tall.

155. Tobacco processed by the Imperial Tobacco Company of Canada Limited was packaged in this Taxi crimp-cut tin. The graphics make it extremely collectible and it is rare and valuable. As more and more American collectors become interested in tin containers, there are less of them to be found in the United States. Canada, the closest neighbor, has recently become an additional hunting ground. The tin is 4.3 inches (10.9 cm) tall.

155

PRESS HERE

TAXI

CRIMP CUT
TOBACCO

156-158. Cigar tins, like tobacco lunch pails, are favored by collectors because they are colorful. The Orcico container in photograph 158 is lithographed in nine colors. Heekin in Cincinnati, Ohio, made it and also the Blue Jay container in photograph 156. The Apache Trail container (photograph 157), also very beautiful, was made by Liberty Can Company of Lancaster, Pennsylvania. They are all 5.5 inches (14.0 cm) tall.

159-162. Collectors have a natural affinity for sets. Each of these four Honey Moon tobacco pocket tins, made during the second decade of the twentieth century, differs slightly from the others. There is also an earlier paper package.

163. This All Nations tobacco tin is another of the so-called flat pocket tins. The container was manufactured by Hasker and Marcuse in the late 1890s and is highly prized because of its extremely colorful graphics. It measures 4 inches (10.2 cm) by 2.75 inches (7.0 cm).

164. This Every Day tobacco tin is one of the most attractive tins ever produced. The subtle colors make it unusual; tobacco labels are more often in primary hues. It was made by S.A. Ilsley and Company of Brooklyn, New York, probably in the 1890s. Tobacco-tin enthusiasts are particularly fond of the legend, "I need thee every day," which apppears on the lower right-hand corner of the tin. It is 4.5 inches (11.4 cm) long and 2.75 inches (7.0 cm) wide.

The Tin Can Book

163

164

THE TIN CAN BOOK

165

Miranda's Dream — Ashes of Am·Bar

166

"Meine Kleine" MANOLI — 10 Stck. Cigarettes

165, 166. Though there are not many to be found, metal cigarette boxes are becoming more and more popular as tobacco-tin collectibles. The designs are quite different from those on other tobacco containers: cigarette tins tend to feature Asian scenes and exotic situations rather than the more down-to-earth images on other tobacco containers. Each of these cigarette boxes features a woman. Miranda's Dream, an English tin, is the earlier of the two. It was probably made between 1915 and 1925. Meine Kleine, a German tin, was made between 1920 and 1930.

ROLY POLY TOBACCO TINS

Roly Poly has become the common name for any one of a group of round tobacco tins lithographed to resemble human beings. (They should be called Brownies, for they were designed to have a secondary use as brownie containers.) Nearly all tin container collectors, whatever their individual specialities, are interested in them.

One reason for the popularity of the Roly Polys is that they belong to a set. There are six different characters, and the tins were made to contain four separate brands of tobacco. One tin of each character would make up a set of six, but a truly full set would include all possible tins made for all four brands of tobacco. A full set, however, would not be composed of twenty-four tins, but only of eighteen. Mayo and Dixie Queen tobacco came in all six tins, but Red Indian and U.S. Marine tobacco came in three characters each. But a set of six is hard enough to find in good condition. The little people have been known to command prices as high as five hundred dollars apiece — a far cry from the fifteen-dollar price tag of just nine years ago.

The Roly Polys have additional appeal because for many years their maker was a mystery. There are no manufacturer's stamps on the tins. An investigation of patent files disclosed the secret: the Roly Polys were made by the Tin Decorating Company (Tindeco) of Baltimore, Maryland, incorporated in 1912, and in operation as early as 1901.

THE TIN CAN BOOK

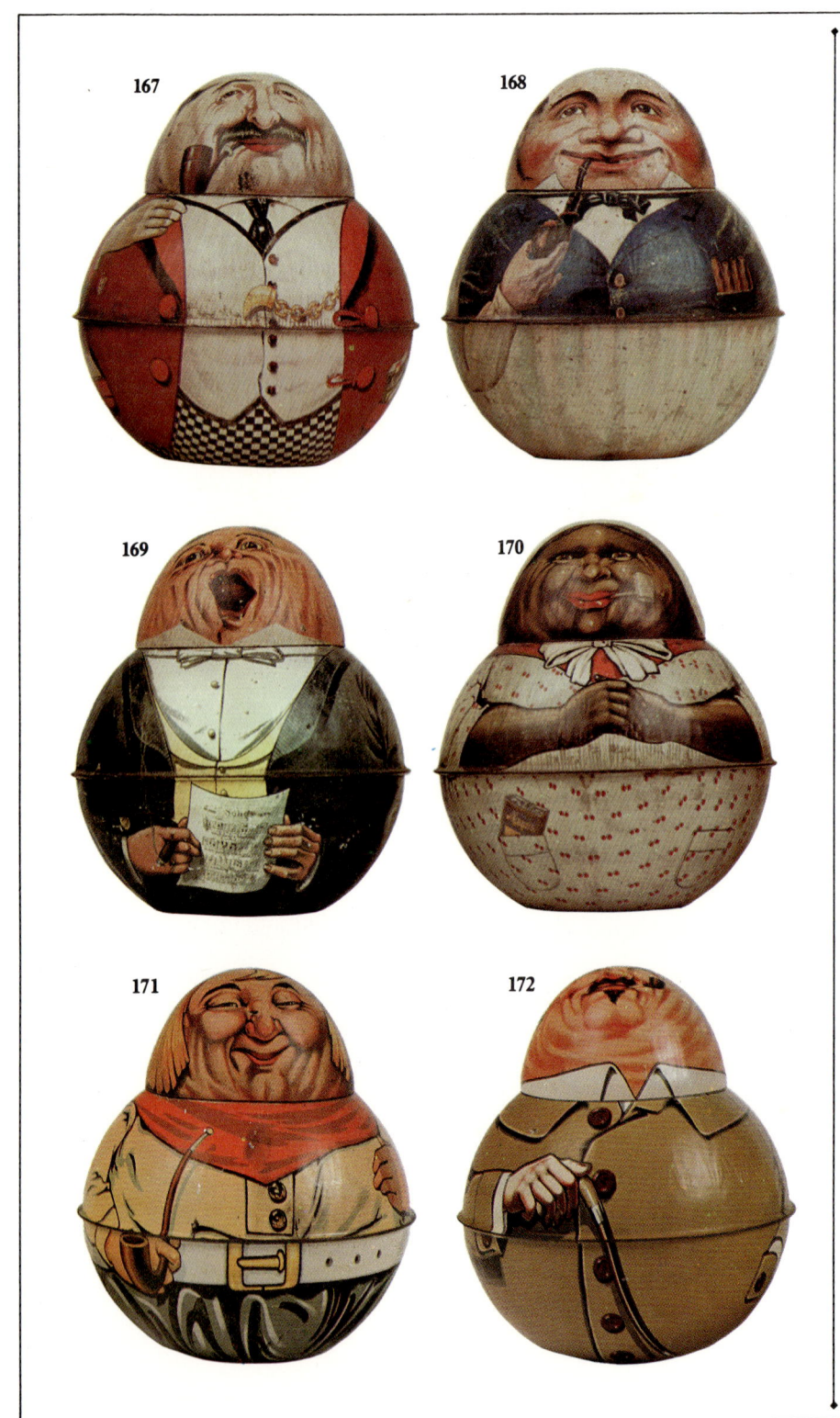

167-172. The six Roly Poly designs: 167. Satisfied customer, 168. Storekeeper, 169. Singing waiter; 170. Mammy; 171. Dutchman; 172. Man from Scotland Yard (or Inspector). The three tins in photographs 167, 171 and 172, particularly Scotland Yard and Satisfied Customer, are rarest. There were fewer Scotland Yard, Satisfied Customer, and Dutchman tins made because these three figures were not made for Red Indian and U.S. Marine tobacco; all six contained Mayo and Dixie Queen tobacco. Also, Scotland Yard, Satisfied Customer and Dutchman were the last of the series to be manufactured. It is thought that the popularity of the tins decreased with time, and fewer of the later designs were made.

173. A little tobacco package was printed on the back of some of the Roly Poly tins to identify the brand. The same package is visible in the apron pocket of the Mammy figure.

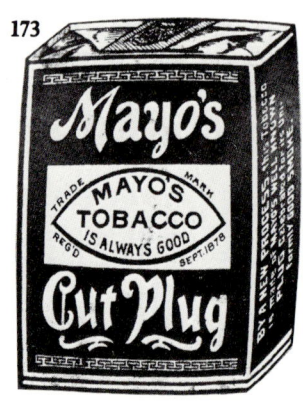

104

174, 175. Patent file searches can be very productive for tin-container collectors. The Roly Poly tins have no manufacturers' mark, and for many years it was assumed they were made by American Can, because the firm sometimes did not mark its tins. But the shape of the container, as well as all six designs, was patented, and patent information showed that they were made by Tindeco for American Tobacco Company. American Tobacco was a conglomerate which controlled Tindeco as well as all four brands of tobacco for which Roly Poly containers were made.

174

DESIGN.
W. I. TUTTLE.
CAN OR CONTAINER.
APPLICATION FILED AUG. 1, 1912.

43,239.

Patented Nov. 5, 1912.

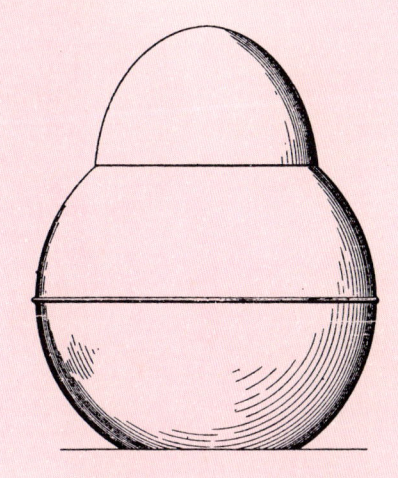

Inventor
Washington I. Tuttle
By Meyers, Cushman & Rea
Attorneys

Witnesses
H. A. Bolnette
G. M. Stucker

COLUMBIA PLANOGRAPH CO., WASHINGTON, D. C.

UNITED STATES PATENT OFFICE.

WASHINGTON I. TUTTLE, OF BALTIMORE, MARYLAND, ASSIGNOR TO AMERICAN TOBACCO COMPANY, OF NEW YORK, N. Y., A CORPORATION OF NEW JERSEY.

DESIGN FOR A CAN OR CONTAINER.

43,239.   Specification for Design.   Patented Nov. 5, 1912.

Application filed August 1, 1912. Serial No. 712,707. Term of patent 14 years.

To all whom it may concern:

Be it known that I, WASHINGTON I. TUTTLE, a citizen of the United States, residing at Baltimore, State of Maryland, have invented a new, original, and ornamental Design for Cans or Containers, of which the following is a specification, reference being had to the accompanying drawing, forming part thereof.

The figure is a front elevation of a can or container, showing my new design.

I claim:

The ornamental design for a can or container, as shown.

WASHINGTON I. TUTTLE.

Witnesses:
JASPER M. BERRY, Jr.,
M. H. CONNOR.

Copies of this patent may be obtained for five cents each, by addressing the "Commissioner of Patents, Washington, D. C."

175

DESIGN.

M. V. TESSIER.

CAN OR CONTAINER.

APPLICATION FILED OCT. 23, 1912.

43,445.

Patented Jan. 7, 1913.

Fig 1

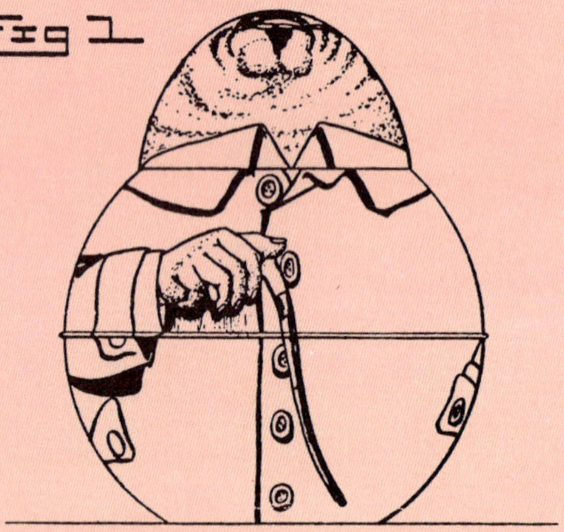

Fig 2

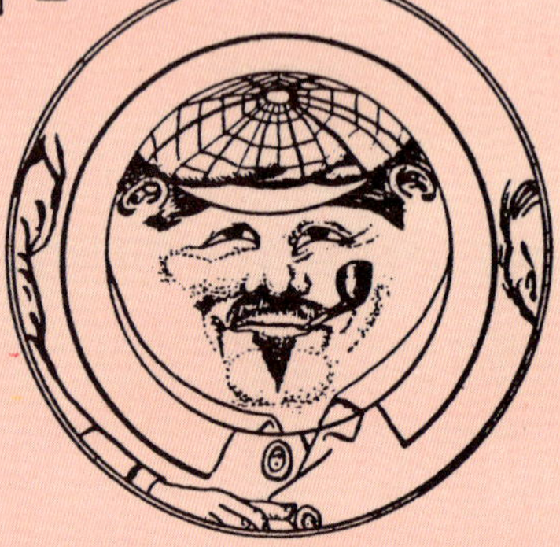

Inventor

Mortimer V. Tessier

Witnesses

H. C. Robinette

Gertrude M. Stucker.

By *Meyers Cushman & Rea*

Attorney

COLUMBIA PLANOGRAPH CO., WASHINGTON, D. C.

SAMPLE TINS

Sample tins or trial sizes, which were handed out free or sold for a nominal amount, have been manufactured nearly as long as full-sized containers. The tiny tins—some as small as a nickel—have the charm of all miniatures. Many are fascinating imitations of their larger counterparts. Of course, some products were always packaged in very small tins. For this reason, collectors look for the word "sample" or "trial size" printed on the side of the package.

Unlike full-sized containers, sample tins were designed to be throw-away items. Nevertheless, they contained a product, however minimal the amount. Many were thrown away, but some were likely to be tossed aside or left in a drawer with their contents intact. (Collectors often find sample tins with the remnants of the product still inside.) As a result, sample tins which have survived are extremely rare but generally in exceptionally good condition.

The tiny tins may be collected to complement a larger collection. For example, sample cocoas or talcum powders may be displayed with matching, full-sized containers. But samples are also collected for themselves. They have the advantage of size—even a large collection fits easily on a few narrow shelves. They are also quite valuable; a good-sized collection may be worth several thousand dollars.

THE TIN CAN BOOK

176   PIQUANTE FACE POWDER

177   DREAMERIE FACE POWDER

178   Complexion Cream Powder de Meridor

179   Encharma Cold Cream Complexion Powder

180   Armand Cold Cream Powder

181   TANGEE FACE POWDER — RACHEL

182

183   OUTDOOR GIRL THE OLIVE OIL FACE POWDER

108

THE TIN CAN BOOK

176-183. A collection of tiny purse-sized cold-cream and powder tins: although they are all elaborately lithographed, they measure only about 1.5 inches (3.8 cm) across. Some collectors concentrate on cosmetic samples.

184-186. Tiny tobacco containers made in the 1920s and 1930s were intended as complimentary packages. These are only 3 inches (7.6 cm) tall and 2 inches (5.1 cm) wide.

187-189. There were samples of household products as well as those for personal use. On the far left is Whiz Soap for "hand or household use" made in the 1920s. It is 1.25 inches (3.2 cm) tall and 1.5 inches (3.8 cm) in diameter. Beside it is a sample of Solarine metal polish for silver, brass, nickel, or tin, made between 1910 and 1920. It is 1.75 inches (4.5 cm) tall. To the right is Putz cream metal polish, from 1900. The tin is nearly 1.75 inches (4.6 cm) tall.

190-192. These three tiny talcum powder samples are decorated in an Oriental style. They are each 2 inches (5.1 cm) tall. Such tins would compose a collection within a collection. The larger collection would be sample tins in general, the middle collection sample talcum-powder tins, and the final sub-collection Oriental sample talcum-powder tins.

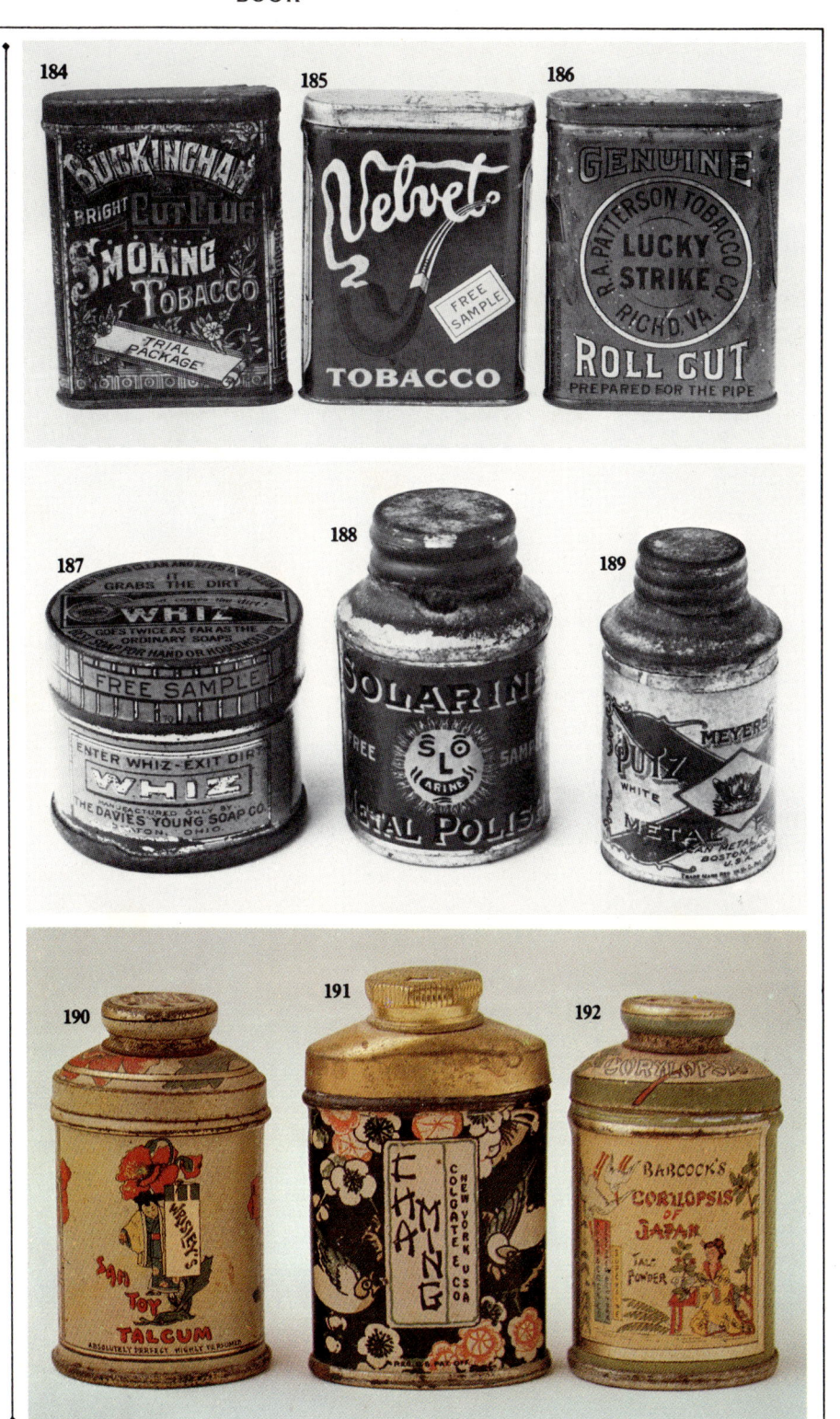

109

THE TIN CAN BOOK

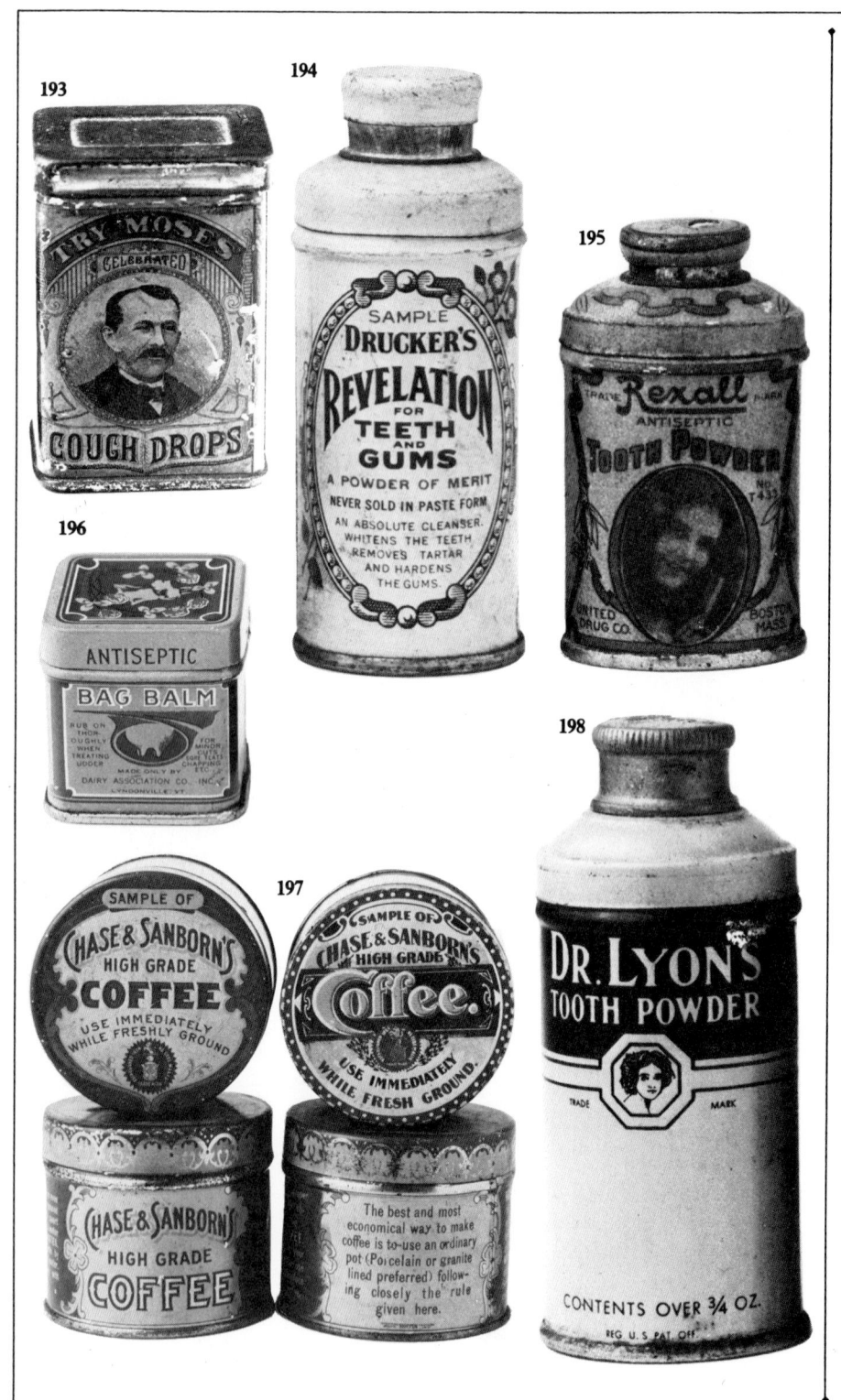

193. Moses Cough drops date from 1875 to 1885. The tins were made by Somers Brothers. This sample tin in the standard orange color is 2.5 inches (6.4 cm) tall and 1.5 inches (3.8 cm) wide. The graphics which make Moses cough-drop tins popular are every bit as clear on this tiny tin as they are on the full-sized container in photograph 122. Sample collectors are fascinated by the extensive detail.

194, 195, 198. All three of these sample tins originally contained tooth powder. The Drucker's Revelation tin, made by August E. Drucker of San Francisco, California, is 2.6 inches (6.5 cm) tall. The tooth powder in the Dr. Lyons container was produced by the R.L. Walkins Company of New York City, not by Dr. Lyons. The tin, a twentieth century version of an early Dr. Lyon's container, is the largest of the three shown. It is 2.8 inches (7 cm) tall. The Rexall sample was made by the United Drug Company, Boston, Mass. It is the smallest — only 2 inches (5.1 cm) tall.

196. This sample tin of Antiseptic Bag Balm made by the Dairy Association Company of Lyndonville, New York. It was available in 1910 and is 1.5 inches (3.8 cm) square.

197. The directions on these Chase and Sanborn coffee samples are the same as those for camp coffee: "Boil water and coffee five minutes, settle by adding dash of cold water." These tins date from 1900 to 1910. They are 2.3 inches (5.9 cm) tall and 2.4 inches (6.1 cm) in diameter.

THE TIN CAN BOOK

199. Many brands of cocoa were available in sample sizes. Monarch is the tallest of this lot at 3 inches (7.7 cm).

200-203. These little Colgate and Comfort powder tins are only 2.2 inches (5.5 cm) wide and 1.7 inches (4.3 cm) tall, respectively. The Mentholatum tins in photographs 202 and 203 were made "to carry with you." The larger one is 1.5 inches (4 cm) in diameter; the smaller 1.4 inches (3.7 cm) in diameter.

THE TIN CAN BOOK

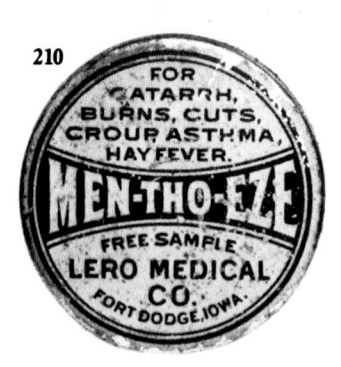

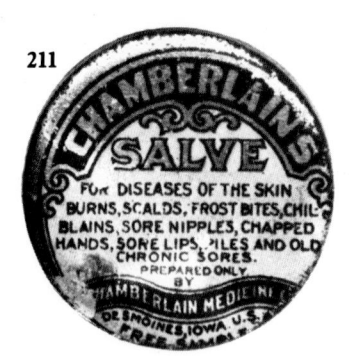

204-211. All these tiny salve and ointment tins were made before 1900. Seamless tins like these were first manufactured in the 1850s, and pharmacists were quick to use them.

204. Milkweed Cream — Sample Box for Tan, Freckles, Sunburn, Redness, Roughness, Blemishes. For Skin Imperfections.

205. Free Sample of Peterson's Ointment — For Itching of Eczema, Blind, Bleeding, Itching and Protruding Piles and Old Sores of Long Standing.

206. Kondon's Catarrhal Jelly — An Instant Relief and Final Cure For Catarrh, Cold in Head, Hay Fever. Apply in Each Nostril 3 Times Daily. Kondon Mfg. Co., Sole Proprietors, Minneapolis, Minn.

207. Bunsens Catarrh Cure — For Catarrh, Catarrhal Deafness, Cold in Head and Hay Fever. Apply in Each Nostril 3 Times Daily. C.W. Beggs, Sons & Co. Price 25¢ & 50¢. Chicago, U.S.A.

208. Dr. Shoop's Green Salve — Makes Lips and Skin Like Velvet. For Cuts, Burns, Bruises, Chilblains, Eczema and All Skin Diseases. Large Jar 25¢. Druggists. From the Dr. Shoop Laboratories, Racine.

209. Pilease — Stops Itching and Allays Irritation Instantaneously. Trade Mark. A Remedy For Itching & Inflamed Piles. Kondon Mfg. Co., Sole Proprietors, Minneapolis, Minn.

210. Men-Tho-Eze — For Catarrh, Burns, Cuts, Croup, Asthma, Hay Fever. Free Sample. Lero Medical Co., Fort Dodge, Iowa.

211. Chamberlain's Salve — For Diseases of the Skin, Burns, Scalds, Frost Bites, Chilblains, Sore Nipples, Chapped Hands, Sore Lips, Piles and Old Chronic Sores. Prepared Only by Chamberlain Medicine Co., Des Moines, Iowa, U.S.A. Free Sample.

STORY TINS

A story tin is any tin container with a rhyme or story printed on it. The original contents of the tin are unimportant; only the exterior is of interest. Various types were produced throughout the collectible era. Some are among the oldest surviving printed tins and all story tins are quite rare. In the early days of tin printing, rhymes and stories were one way to decorate a tin. Colorful characters and situations offered exciting graphics, and the lack of product identification provided a package which could contain several different products. These tins continued to be produced even after color was widely used in tin printing.

The rhymes on the early tins were not associated with children, but the majority of later rhyming tins were produced with youngsters in mind. Huntley and Palmers' biscuit tins were sometimes decorated with nursery rhymes and story-book characters, and there were several series of rhyming tins made in the United States in the 1920s and 1930s.

Although they are not specifically story tins (because they have no printed words), the Harrison Cady tins have been included in this section. Harrison Cady was an illustrator in the 1920s who became famous for his drawings for children.

Story tins may be a part of another collection, an English biscuit tin or pre-twentieth century tin collection, for example. They may also be collected for themselves.

THE TIN CAN BOOK

212

213

212. All of the known Peter Rabbit tins are thought to have been made by the Baltimore tinmaker, Tindeco, a wholly-owned subsidiary of the American Tobacco Company known for its tobacco tins, which also produced a large line of non-tobacco items. The Peter Rabbit tins date from 1924 and include: a candy canister; three candy lunch pails; three tin plates, a tin tray and tin Easter egg; an Easter candy pail; talcum powder tins; a peanut butter pail and tin cups. Magazine illustrator Harrison Cady drew the designs for the tins. He was a prolific and successful artist who drew a series of Peter Rabbit cartoons for the Sunday comic section of the "New York Herald Tribune" and contributed many children's stories to the "Ladies Home Journal" and "Good Housekeeping" magazines.

213. Lovell and Covel candy pails were made in the 1920s and 1930s. They were small—3 inches (7.6 cm)—and contained only three ounces of candy. The manufacturers were more interested in selling candy than in being accurate: Red Riding Hood pacifies the world with candy, and Peter Rabbit steals candy instead of carrots.

214-219. This rhyming story tin was made by McCall and Stephen Biscuit Manufacturers of Glasgow, Scotland. Katie is apparently a willing lass, and the biscuit tin is quite pretty. It is unusual both because it is an adult tin, and because it is printed in color. It is 4.75 inches (12.1 cm) tall.

115

220-226. This six-sided Froggy Went A' Courtin' tin is prized for its shape, its age and its rhyme. The tin is quite rare. It was made in England by Huntley Boorne and Stevens in the 1880s. Such tins as this one and the Heathen Chinee tin (page 118) were made as tea containers. The Froggy Went A' Courtin' tin is 6.5 inches (16.5 cm) tall.

A Frog he would a-wooing go,
Heigho, says Rowley!
Whether his mother would let him or no.
Heigho, says Antony Rowley!

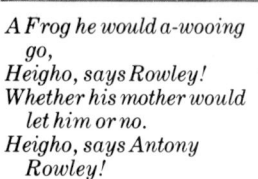

So off he set with his opera hat,
Heigho, says Rowley!
And on the road he met with a rat.
Heigho, says Antony Rowley!

Pray, Mrs. Mouse, will you give us some beer?
Heigho, says Rowley!
For Froggy and I are fond of good cheer.
Heigho, says Antony Rowley!

THE TIN CAN BOOK

*Since you have caught cold
Mr. Frog, Mousey said,
Heigho, says Rowley!
I'll sing you a song that I
have just made.
Heigho, says Antony
Rowley!*

*But while they were all
a-merry making,
A Cat & her Kittens came
tumbling in.
The Cat she seized the Rat
by the crown;
The Kittens they pulled the
little Mouse down.*

*This put Mr. Frog in a
terrible fright.
He took up his hat and
wished them good night.
But as Froggy was crossing
over a brook
A lily white Duck came &
gobbled him up.
Heigho, says Antony
Rowley!*

Around the top of the can:

*So there was an end of one,
two, and three
Heigho, says Rowley!
The Cat, the Mouse and the
little Froggy
Heigho, says Antony
Rowley!*

117

THE TIN CAN BOOK

227. WHICH, I WISH TO REMARK,—
AND MY LANGUAGE IS PLAIN,—
THAT FOR WAYS THAT ARE DARK
AND FOR TRICKS THAT ARE VAIN,
THE HEATHEN CHINEE IS PECULIAR,
WHICH THE SAME I WOULD RISE TO EXPLAIN.

228. BUT THE HANDS THAT WERE PLAYED
BY THAT HEATHEN CHINEE,
AND THE POINTS THAT HE MADE
WERE QUITE FRIGHTFUL TO SEE,
TILL AT LAST HE PUT DOWN A RIGHT BOWER
WHICH THE SAME NYE HAD DEALT UNTO ME.

229. THEN I LOOKED UP AT NYE,
AND HE GAZED UPON ME,
AND HE ROSE WITH A SIGH,
AND SAID, "CAN THIS BE?
WE ARE RUINED BY CHINESE CHEAP LABOR,"
AND HE WENT FOR THAT HEATHEN CHINEE.

230. IN HIS SLEEVES, WHICH WERE LONG,
HE HAD TWENTY-FOUR PACKS,—
WHICH WAS COMING IT STRONG,
YET I STATE BUT THE FACTS:
AND WE FOUND ON HIS NAILS, WHICH WERE TAPER,
WHAT IS FREQUENT IN TAPERS,—THAT'S WAX.

227-231. The story of the Heathen Chinee, depicted on this Ginna and Company tea canister from the 1870s or 1880s, illustrates a deep American prejudice of the era: the narrator laments the heavy influx of Chinese labor into the United States. He is fearful that domestic unemployment will result. Of course, such threatening foreigners must be card sharps. The tin is 5 inches (12.7 cm) tall.

118

BEER CANS

Beer cans can be very valuable and some have been known to sell for more than one hundred dollars each. However, the BCCA (Beer Can Collectors of America) suggests that beer cans only be traded for other beer cans. The club does not allow selling during its functions. Thousands of collectors belong to BCCA, and a national magazine recently estimated that there are some five-hundred thousand beer can collectors in the United States. If the figure is correct, beer can collecting must be one of the most attractive hobbies in the country.

Like other tin containers, beer cans are collectible if they are in good condition. A can may also have value if it is rare. Collectors are always on the lookout for labels that were test-marketed but never sold nationally, color changes in a label from one printing to the next, off-beat brand names sold only briefly or cans produced in small quantities for out-of-the-way breweries. Very old cans in good condition are also rare.

Beer cans currently available are as collectible as cans of the past. Collectors scour local shops for special-edition cans, and cans from distant locations are prized. In order to preserve the tab-tops of contemporary containers, collectors open them from the bottom.

The Tin Can Book

232
PLAY MATE malt liquor — CONTENTS 12 FL. OZ.

233
PLAY MATE PREMIUM BEER with the zing of malt liquor — CONTENTS 12 FL. OZ.

234
Mitchell's Premium Beer — THE BALANCED BREW — Net Contents 12 Fluid Ounces

235
Mitchell's Premium Beer — THE BALANCED BREW — Mitchell Brewing Co., El Paso, Texas

236

THE TIN CAN BOOK

232, 233. These Playmate beer cans were manufactured only for a short time. Legend has it that the brewery was sued for using the name. The beer can is more rare than the malt-liquor can, because fewer were made.

234-236. The most minute differences in a label can make one beer can more rare than another. The two Mitchell's premium beer cans in photographs 234 and 235 appear identical in every respect. But in the loop of the "M" on the bottom of the can in photograph 235 the word "Harry" is written (see detail in photograph 236). It is not known why Harry is not mentioned on the other can.

237. This Fox Head ale can was made in the late 1950s. It is not particularly rare, but it is collectible simply because it is good-looking.

238. The reverse situation is true of this 7-Eleven beer can. It is not really very attractive at all. It is, however, very rare; although there are various different 7-Eleven cans, this is the only one of its kind known.

239. This Red Velvet ale can was made in the late 1950s for the Standard Brewing Company of Cleveland, Ohio. It is extremely rare.

240. An Acme beer can is not rare. It was, however, recently chosen as the BCCA Beer Can of the Year. Since modern cans like this one are made using an extrusion process, the label is printed on after the can is formed. At first, can makers had some trouble with smearing on the edges. The technique, however, is being perfected.

237

238

239

240

121

241-242. There is one known atavism in the beer-can world. The Milwaukee Brewing company of Hammonton, New Jersey, has recently produced a new cone-top can.

243-245. Soul malt-liquor and beer were made by the Maier Brewing Company of Los Angeles, California, in the late 1960s. A 12-ounce can was also made for malt-liquor, but it never held liquid. Technically, such a can would not be collectible, though beer-can collectors often include mugs and demonstration cans in their collections.

246. These 007 beer cans, like the Soul cans and Playmate cans, belong to what is known as the "popular-rare" group. They are not easy to find, but there are enough of them around to make them recognizable as rare. The truly rare beer cans are too rare to be well-known. Beer cans can command a high price; a set of 007 cans have been known to sell for eight hundred dollars.

247. Crown Cork and Seal Company introduced its own version of the cone-top can in the early 1940s. Its drawn-metal "Crowntainer" was constructed using a seamless body, a bottom, and a cap. Crowntainers were shorter than cone-top cans made by other manufacturers. Standard cone-top cans measure 5.5 inches (14.0 cm); drawn-metal cone-top cans measure 5 inches (12.7 cm) and flattop cans measure 4.5 inches (11.4 cm).

248. There are two basic styles of construction in cone-top beer cans. This is the older of the two. Continental Can brought out the first cone-top can in this design in late 1935. The cans are made in three pieces; the body was formed and seamed, then the cone and bottom were added. It ultimately lost the race with the flat-top can, and was obsolete by the 1950s.

THE TIN CAN BOOK

246

247

248

249, 250. Camden County Beverage Company was briefly taken over by Esslinger's, Inc.: the can on the left reads, "Distributed by Camden County Beverage Company, N.J., Brewed and Packaged by Esslinger's Inc., Phila., Pa."; and the can on the right reads, "Brewed and Packed by Camden County Beverage Company, Camden, N.J., New York Sales Office, 1 Park Place, New York, N.Y." Beer-can collectors look for such small differences. The can on the left was produced only for a very short time and is therefore rare.

251, 252. When flat-top beer cans were first introduced in 1935, retailers included a free can opener with every purchase. For nearly ten years instructions on how to open flat-top cans were printed on the labels.

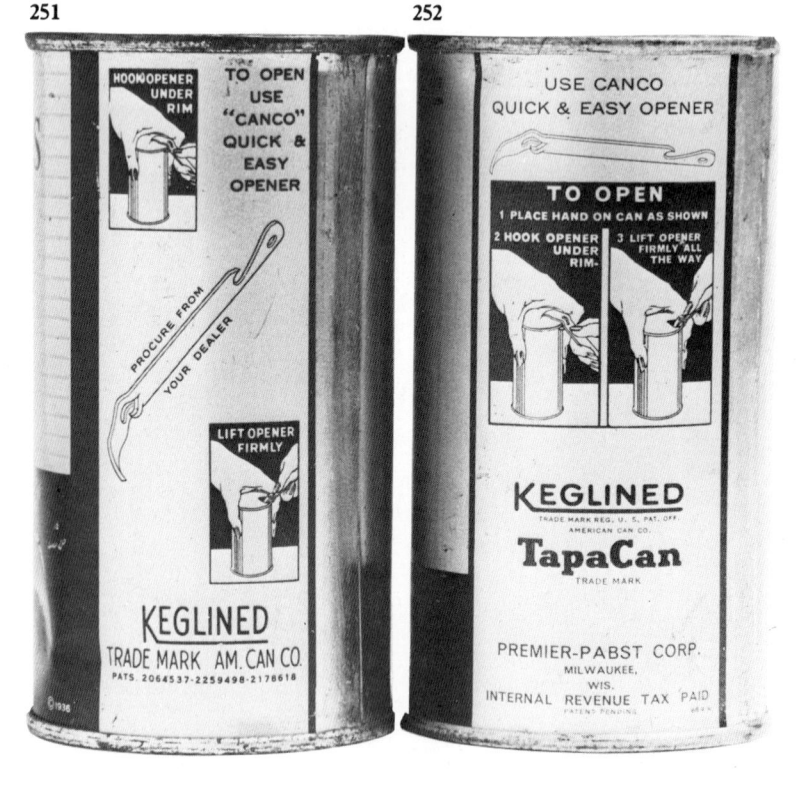

253, 254. Half-size beer cans such as these were first available in the late 1950s. Collectors of beer cans focus on size as well as design. Small, half-size cans and big, gallon containers are equally collectible.

255. All eight of these beer cans are rare in some way. Milwaukee Valley beer cans, for instance, were made only once, in 1959, for a Wisconsin brewer, but were sold only in California. Pickwick brew is also unique; it was a combination of lager beer, malt-liquor, and ale. The Rheingold can features the winner of the 1957 Miss Rheingold contest. All three contestants and the winner were pictured on the brewery's labels. Triple Crown is Ballantine's rarest label. It was test-marketed in Baltimore, Maryland, in 1967, but never sold nationally. The Gretz can pictured here is one of a series of cans depicting different cars made in the 1950s. The sets are highly prized. The State Fair label is white with a blue and red logo. The white label was printed only for a short time in 1964 before the more common silver label was used. Eastern beer was made in the Midwest to be sold on the West Coast. This particular label, featuring an urban skyline, is hard to find. Bohio beer was made in Camden, New Jersey, in the 1950s. Because the beer was only sold locally, few of the cans are available for collectors.

125

THE TIN CAN BOOK

256, 257. Currently available foreign beer cans are quite collectible because they are harder to obtain than American cans. Difficulty in procurement is as important in beer-can collecting as actual rarity. These Japanese cans are particularly colorful. The two cans on the far left were made for Asahi lager beer; the other six for Suntory beer. Swedish 16-ounce cans and German gallon kegs are also hunted by beer-can collectors.

THE TIN CAN BOOK

257

THE TIN CAN BOOK

258, 259. Beer cans also have their legendary histories. West Arm and Deep Cove cans commemorate the construction of a large hydro project in New Zealand. Deep Cove was the southern terminal, and West Arm the northern. It was a long and arduous task: construction began in 1963 and the project was not completed until 1971. Social clubs evolved at both ends, and a rivalry grew up between them. In late 1967, beer was canned under the Deep Cove label for the southern clubs, and by early 1968, West Arm beer was also available. Each of the labels was manufactured for about one year. They are highly collectible because the cans were only made for a short time, and because they have an interesting history.

260. This Gunther all-aluminum beer can is quite rare. It was first manufactured in 1959, at the same time Coors introduced its all-aluminum can. Like most very rare beer cans, the Gunther can is not well-known. Like the Coors can, the Gunther can contained only seven ounces of beer.

261. Hawaii Brewing Corporation Limited of Honolulu, Hawaii, introduced an all-aluminum can in 1958, the year before Gunther and Coors came out with their versions. The Primo beer can is full-sized and features an outstanding paper label. The clarity of the blue and red graphics against the silver can makes it a favorite.